Student Handbook
for
Discrete Mathematics
with Ducks: SRRSLEH

Student Handbook

for

Discrete Mathematics

with Ducks: SRRSLEH

Student Reference, Review, Supplemental Learning, and Example Handbook

sarah-marie belcastro

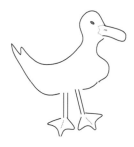

serious mathematics treated with levity

all. the. time. srsly. srrsleh?

CRC Press
Taylor & Francis Group
Boca Raton London New York

CRC Press is an imprint of the
Taylor & Francis Group, an **informa** business

AN A K PETERS BOOK

CRC Press
Taylor & Francis Group
6000 Broken Sound Parkway NW, Suite 300
Boca Raton, FL 33487-2742

First issued in hardback 2017

© 2016 by Taylor & Francis Group, LLC
CRC Press is an imprint of Taylor & Francis Group, an Informa business

No claim to original U.S. Government works

Version Date: 20150410

ISBN-13: 978-1-4987-1404-4 (pbk)
ISBN-13: 978-1-138-43585-8 (hbk)

Library of Congress Cataloging-in-Publication Data

Belcastro, Sarah-Marie.
    Student handbook for Discrete mathematics with ducks : SRRSLEH : student reference, review, supplemental learning, and example handbook / Sarah-Marie Belcastro.
        pages cm
    "An A.K. Peters Book."
    "Serious mathematics treated with levity."
    ISBN 978-1-4987-1404-4 (alk. paper)
    1. Mathematics--Textbooks. 2. Mathematics--Problems, exercises, etc. 3. Computer science--Mathematics--Textbooks. 4. Computer science--Mathematics--Problems, exercises, etc. I. Title. II. Title: Discrete mathematics with ducks. Student handbook.

QA39.3.B386 2015
510--dc23   2015010898

Visit the Taylor & Francis Web site at
http://www.taylorandfrancis.com

and the CRC Press Web site at
http://www.crcpress.com

Deduckation: In memory of Klaus Peters (1937–2014), without whose many influences on the publishing world none of the books I've contributed to would exist.

Contents

Preface for ... Everyone!

1 What's *This* All About?

Once upon a time, I wrote a textbook (you might have heard of it; it's called *Discrete Mathematics with Ducks*). A couple of years later, I tried to look something up in that textbook. Wow! What a pain! Even though the glossary told me on what page I'd put a definition, I couldn't quickly pick out that definition from the wall of text. What was I expecting students to do?

On reflection, I remembered that I was expecting students to be introduced to material from a particular perspective when reading the text. That is, I wanted students to *learn* from the text, and that's not the same as hoping readers can easily *refer* to items within the text. But of course it is a reasonable expectation for a reader to be able to find information easily. Thus I concluded I needed to write some kind of reference.

Why, you might wonder, did I not simply revise the textbook itself in order to make it work as a reference as well? There are a lot of good reasons the textbook is written in narrative form:

- It allows for fuller explanations.

- It is colloquial and thus less intimidating.

- It places material in context rather than isolating it.

- It models the way mathematicians think about mathematics instead of the way they formally write mathematics.

- It provides practice for students in abstracting mathematical content.

For these reasons, I concluded that a separate volume was necessary.

On top of the whole I-can't-look-up-definitions-in-my-own-textbook thing, I had noticed that there were multiple reviews on amazon.com—both positive and negative—that pointed out the lack of reasonably challenging solved problems in *Discrete Mathematics with Ducks*. A student could easily master problems at the Check-Yourself level, but get stuck (especially when using the book for self-study) on harder problems. Nontrivial problems with accompanying solutions would help with that. (It later occurred to me that students using other textbooks might find this useful as well!)

I didn't want to simply give solutions to the homework problems already in the textbook, because there still needed to be problems instructors could assign as homework. In other words, I needed to write more problems. And I didn't want to include those in the main text, because it's already pretty long. So, I *really* needed to make a separate document.

Thus the idea for the *Student Reference, Review, Supplemental Learning, and Example Handbook* (henceforth SRRSLEH, pronounced "seriously") was born. (While we're at it, henceforth *Discrete Mathematics with Ducks* will be DMwD.)

2   How SRRSLEH Is Organized, and How You Should Use It

Every chapter in SRRSLEH is matched to the corresponding chapter of DMwD. It has

- 🦆 a list of the notation introduced in the corresponding DMwD chapter,

- 🦆 a list of definitions students need to know from the corresponding DMwD chapter,

- 🦆 theorems/facts of note appearing in the corresponding chapter,

- 🦆 a list of proof techniques introduced, with templates and/or examples given for each one,

- 🦆 a selection of the examples from DMwD, written out formally and briefly rather than colloquially or with stream-of-consciousness,

🐦 new examples,

🐦 a set of new, non-easy problems, and

🐦 a set of solutions to those new, non-easy problems. Some solutions are marked as being models for hand-in-worthy writing.

If you are using SRRSLEH in conjunction with a traditional discrete mathematics textbook, this organization should enable you to figure out how the chapters in this book correspond to the chapters in your textbook.

If you are using SRRSLEH with DMwD, please notice that SRRSLEH is no substitute for DMwD. It is important to read DMwD *first* because that's where all the explanations are—DMwD will walk you through how to think about each topic, and it helps you to see the mathematics underneath a practical (or silly) problem setup. Plus, there is a lot of material that you can, and should, see for yourself. DMwD gives you the tools to work through that mathematics on your own and with your peers; don't spoil the fun or exempt yourself from the mental exercise by looking in SRRSLEH first. Seriously (not just SRRSLEH), you get better at math by practicing it, and trying the exercises in DMwD gives you that practice. If you haven't looked in DMwD yet, go read the *Preface for Students and Other Learners* right now.

SRRSLEH is for summarization and review of material. (For detailed review, you will still want to go back to DMwD.) Read it over a week after you've covered the corresponding DMwD chapter in class. It will remind you of which terms you're supposed to know and what results you should keep in mind. Some examples that were treated in exquisite detail in DMwD are written short-and-sweet in SRRSLEH so you don't have to wade through tons of text to pick out the ideas for a quick refresher. And SRRSLEH is here to give you additional practice with the material—look at all those new problems for which you can check your answers!

So, when you can't remember a definition, go to SRRSLEH. What did that one theorem say? SRRSLEH will tell you. If you want the *proof* of the theorem, go to DMwD for the full story. It's even possible that if you're stuck on your homework, there will be a solved problem in SRRSLEH that will help.

Finally, one question and one answer:

Question. How did you choose which examples from DMwD to rewrite in SRRSLEH?

Answer. If an example was written with a wack story or extended explanation, it's been rewritten here without the wack story or extended explanation. But if it's the mathematics that was convoluted, and the story was plain, then rewriting wouldn't help; nothing will help you other than re-reading (sorry).

Acknowledgments

Thanks still go out to all the people and ducks who inspired and aided the writing of *Discrete Mathematics with Ducks*; see that volume for details. (For those keeping track, Adam M. and I are friends again.)

My new gratitude: Thanks to the reviewers on Amazon who *didn't* like *Discrete Mathematics with Ducks*—your one-star reviews helped me see what was needed in this supplement. And thanks to Klaus Peters, sadly now deceased, who encouraged me to continue enhancing DMwD. I'm grateful to my ITFF peeps for providing inspiration for Problem 1 in Chapter 1 and Problems 2 and 4 in Chapter 5. Ben Eisen, thanks for sending me the *Elder Sign* dice info. Thanks to Tom Hull for many fruitful conversations about writing problems throughout the text, though particularly for conversations about Chapters 4, 11, and 13. Finally, I'm thankful to my personal feline Pantalaimon, who sat next to me during much of the work on SRRSLEH.

Credit where credit is due: All new figures were made in EazyDraw, a technical illustration program for the Macintosh. Dave Mattson, the creator, gives excellent and prompt user support.

Problem 9 in Chapter 1 was suggested by Doug Shaw.

Problem 10 in Chapter 5 includes a quote from Jane Austen.

Problem 3 in Chapter 7 was inspired by an *Art of Problem Solving* student whose username is ninjataco.

Problem 1 in Chapter 5, Problem 1 in Chapter 8, Problem 4 in Chapter 10, Problem 7 in Chapter 12, Problem 3 in Chapter 14, and Problem 10 in Chapter 15 were donated or suggested by Tom Hull.

Part I

Theme: The Basics

Chapter 1 🦆

Counting and Proofs

Content covered in this chapter includes basic counting (the sum, product, and pigeonhole principles) and an introduction to direct proof.

1.1 Chapter 1 Definitions and Notation

1.1.1 Notation

$\{a, fins, 2, 16\}$: The curly brackets indicate that this is a finite set, and the items separated by commas are the elements of the set.

$a \in A$: This means that a is an element of the set A.

(a, b) and (a, b, c): An ordered pair (also an element of the set $A \times B$) and an ordered triple (also an element of the set $A \times B \times C$). This generalizes to an ordered n-tuple.

$A \times B$: The Cartesian product of sets A and B, with elements (a, b) where $a \in A$ and $b \in B$. This is really defined properly in Chapter 2.

$|A|$: The number of elements in a set A.

1.1.2 Definitions

disjoint sets: Sets with no elements in common.

union of sets: The union of two sets A, B (or many sets $A, B, \ldots, N$) is a set containing all members of A and of B (and of $C, \ldots, N$).

subset: A subset A of a set B is a set all of whose members are also members of B.

definition: A precise statement of the meaning of a term. (Think dictionary, but better.)

conjecture: A statement proposed to be true and made on the basis of intuition and/or evidence from examples.

theorem: A statement that can be demonstrated to be true.

proof: A justification of the truth of a statement using reasoning so rigorous that the argument compels assent.

proposition: A smallish theorem, or a theorem offered (proposed) to the reader.

counterexample: A particular case in which a statement is untrue. For example, 3 is a counterexample to the statement *All numbers are even*.

prime number: An integer $n > 1$ is prime if the only positive divisors of n are n and 1.

even number: A number evenly divisible by 2. Equivalently, a number m is even if $m = 2k$ for some integer k.

odd number: A number m is odd if $m = 2k + 1$ for some integer k.

binary number: A number expressed using only the digits 0 and 1, with counting proceeding as $1, 10, 11, 100, 101, 110, \ldots$ and with places representing powers of two, increasing to the left and decreasing to the right.

1.2   Chapter 1 Facts and Theorems

The sum principle. The number of elements in a finite number of disjoint finite sets $A, B, \ldots, N$ is the sum of their sizes $|A| + |B| + \cdots + |N|$.

The product principle. The number of elements in the Cartesian product of a finite number of finite sets $A \times B \times \cdots \times N$ is the product of their sizes $|A| \cdot |B| \cdots \cdots |N|$.

Fact. The same Cartesian product can be grouped as $|B|$ copies of A, or as $|A|$ copies of B.

Theorem 1.5.2. A set with n elements has 2^n subsets.

Figure 1.1. Pigeons sitting in pigeonholes.

Fact. If two sets A and B are in one-to-one correspondence, then they have the same size.

The pigeonhole principle. If you have more pigeons than pigeonholes, then if every pigeon flies into a hole, there must be a hole containing more than one pigeon. (See Figure 1.1.)

The generalized pigeonhole principle. If you have more than k times as many pigeons than pigeonholes, then if every pigeon flies into a hole, there must be a hole containing more than k pigeons.

1.3 Chapter 1 Proof Techniques: Direct Proof and Pigeonhole Tips

Template for a direct proof:

1. Restate the theorem in the form *if (conditions) are true, then (conclusion) is true*. Most, but not all, theorems can be restated this way. (For example, some theorems are secretly phrased as follows: *(conditions) are true if and only if (conclusions) are true.*)

2. On a scratch sheet, write *assume (conditions) are true* or *suppose (conditions) are true*.

3. Take some notes on what it means for (conditions) to be true. See where they lead.

4. Attempt to argue in the direction of *(conclusion) is true*.

5. Repeat attempts until you are successful.

6. Write up the results on a clean sheet, as follows.

- 🐦 Theorem: (State theorem here.)

- 🐦 Proof: Suppose (conditions) are true.

- 🐦 (Explain your reasoning in a logically airtight manner, so that no reader could question your statements.)

- 🐦 Therefore, (conclusion) is true. (Draw a box or checkmark or write Q.E.D.—the abbreviation of *quod erat demonstrandum*, Latin for "which was to be demonstrated"—to indicate that you're done.)

Admittedly, there is a lot of grey area in just how one should argue in the direction of *(conclusion) is true*. This is where the creativity and art of proof come in. One must be careful to avoid the temptation to start with the conclusion and work backwards and then hand that in as a finished proof; the steps have to be reversible and presented in the appropriate order. (One must also avoid the temptation to give a few examples and call it a day. That is *not* a proof.)

How to apply the pigeonhole principle:

1. Figure out what represents the pigeons.

2. Figure out what represents the pigeonholes.

3. Figure out how pigeons correspond to holes.

1.4   Some Straightforward Examples of Chapter 1 Ideas

Example 1.3.2 rewritten.   A meal at the Restaurant Quatre-Étoile consists of an appetizer, a main dish, and a dessert.
The set of appetizers A has 4 members; the set of main dishes M has 5 members; and the set of desserts D has 2 members.
The product principle says that the number of different meals that could be ordered is $|A| \cdot |M| \cdot |D| = 4 \cdot 5 \cdot 2 = 40$.

An example of number types.   Let's look at the numbers $5, 28$, and 10.
Because $5 \cdot 1 = 1 \cdot 5 = 5$ and there are no other integer factorizations of 5, it is prime. It is also odd because $5 = 2 \cdot 2 + 1$.
On the other hand, $28 = 14 \cdot 2$ is even, and not prime; it is also not binary because it uses digits other than 0 and 1. We can rewrite $28 = 16 + 8 + 4 = 2^4 + 2^3 + 2^2$, so its binary representation is 11100.

The number 10 could be the decimal number 10 or the binary number 10 representing the decimal number 2 (which, by the way, is both even and prime).

Example 1.5.4 rewritten. Let's show that in San Francisco, at least four people have the same number of hairs on their heads.

No one has more than 200,000 hairs on her/his head, and there are at least 830,000 people in San Francisco (as of 2013, says Google).

So, there are 200,000 different numbers of hairs, and at least 4 times that many people, so there must be at least 4 people with the same number of head-hairs.

Even if we assigned different hair numbers to the first 200,000 people, and did this again to the second and third 200,000 people, there would still be more people (who would have to quadruple-up on some hair numbers).

Example 1.5.5 rewritten. Let's show that given any list of 25 numbers with five or fewer digits, two subsets of the list have the same sum.

No number can be more than 100,000, so the sum of all 25 of them is less than 2,500,000 and there are no more than 2,500,000 different possible sums the subsets could have.

By Theorem 1.5.2, there are $2^{25} = 33,554,432$ possible subsets.

There are *way* more subsets than sums, so two of the subsets must have the same sum.

1.5   More Problems for Chapter 1

Those solutions that model a formal write-up (such as one might hand in for homework) are to Problems 2, 4, and 6.

1. A Timbuk2 custom messenger bag comes in four sizes, has 46 options for the left-panel and center-panel and right-panel fabrics, 18 different binding options, 27 logo colors, 11 liner colors, three options for pocket style, two handednesses, and 47 different options for the strap pad. (Really, not kidding—these numbers came from the Timbuk2 website in October 2014.) How many different custom messenger bags could one order?

2. Prove that the product of any three odd numbers is also odd.

3. Takeo, a paper store in Tokyo, has walls lined with coded drawers. Each code designates a type of paper. One such drawer is 2Q08. If the first entry has to be 1, 2, or 3 (there are only three walls with drawers), the second is a letter, and the last two are numbers, then how many drawers could Takeo have?

4. You want to buy an electric car. The Chevy Volt comes in eight colors (red, brown, grey, pale blue, two blacks, two whites), offers three kinds of wheels, and has five kinds of interiors (two cloth, three leather). The Tesla comes in nine colors (black, two whites, two greys, brown, red, green, blue), and gives a choice of three roof styles (one is glass), four wheel styles, four seat colors, four dash-board prints, and three door-trim colors. There are three versions of the Nissan Leaf (S, SV, SL), each of which comes in seven colors (two whites, two grays, red, blue, black). How many different choices of car do you have?

5. Prove, or find a counterexample: the sum of two consecutive perfect cubes is odd.

6. How many 4-digit phone extensions have no 0s and begin with 3?

7. In 2012, there were 3,952,841 live births in the US. (Source: http://www.cdc.gov/nchs/fastats/births.htm) Did there have to be two of these births within the same second?

8. How many length-8 binary strings have no 0s in the fourth place?

9. You receive a choose-your-own-adventure certificate for a jewelry store! The deal is that you get to pick one of eight precious gems, and either a ring or a bracelet to put it in. There are three possible ring styles, and six possible bracelet styles.

   (a) How many possible prizes are there?

   (b) How did you answer the previous question? If you used the product principle first, re-answer the question using the sum principle first. (And if you used the sum principle first, re-answer the problem using the product principle first.)

   (c) On closer look, you realize that neither the ruby nor the emerald would look good on the bracelet. How many prizes are still possible?

10. I have a lot of stuff in my stuff-holder: Six ball-point pens, a silver star wand, three teal signature pens, a bronze-yellow colored pencil, five liquid ink pens, three mechanical pencils, a highlighter, six permanent markers, seven gel pens, a Hello Kitty lollipop, two markers, three wooden pencils, a 3-inch-long pen, a calligraphy marker, a pen shaped like a cat, and a pair of left-handed office scissors.

1.6   More Solutions for Chapter 1

1. A Timbuk2 custom messenger bag comes in four sizes, has 46 options for the left-panel and center-panel and right-panel fabrics, 18 different binding options, 27 logo colors, 11 liner colors, three options for pocket style, two handednesses, and 47 different options for the strap pad. (Really, not kidding—these numbers came from the Timbuk2 website in October 2014.) How many different custom messenger bags could one order?

    This is a total mix-and-match situation, so the product principle applies and we multiply together all the numbers of options. There are $4 \cdot 46 \cdot 46 \cdot 46 \cdot 18 \cdot 27 \cdot 11 \cdot 3 \cdot 2 \cdot 47 = 586{,}964{,}112{,}768$ ways to order a Timbuk2 custom messenger bag.

2. Prove that the product of any three odd numbers is also odd.

    We first name the three odd numbers n_1, n_2, n_3.
    Because they are odd, each can be written in the form $2k+1$—but the k is likely different for each, so we have
    $n_1 = 2k_1 + 1, n_2 = 2k_2 + 1$, and $n_3 = 2k_3 + 1$. The product of the numbers is
    $n_1 n_2 n_3 = (2k_1 + 1)(2k_2 + 1)(2k_3 + 1)$. Expanding this expression, we get
    $8n_1 n_2 n_3 + 4n_1 n_2 + 4n_1 n_3 + 4n_2 n_3 + 2n_1 + 2n_2 + 2n_3 + 1$, which can be rewritten as
    $2(n_1 n_2 n_3 + 2n_1 n_2 + 2n_1 n_3 + 2n_2 n_3 + n_1 + n_2 + n_3) + 1 = 2q + 1$ for some integer q.
    Therefore the product of any three odd numbers is also odd.

3. Takeo, a paper store in Tokyo, has walls lined with coded drawers. Each code designates a type of paper. One such drawer is 2Q08. If the first entry has to be 1, 2, or 3 (there are only three walls with drawers), the second is a letter, and the last two are numbers, then how many drawers could Takeo have?

    We think of filling slots: The first slot has 3 possibilities, the second 26, and the third and fourth slots have 10 possibilities each. The product principle says there are $3 \cdot 26 \cdot 10 \cdot 10 = 7800$ paper drawers.
    (It turns out that at Takeo there are no drawers that end in 00, so this is certainly an overestimate.)

4. You want to buy an electric car. The Chevy Volt comes in eight colors (red, brown, grey, pale blue, two blacks, two whites), offers three kinds of wheels, and has five kinds of interiors (two cloth, three leather). The Tesla comes in nine colors (black, two whites, two greys, brown, red, green, blue), and gives a choice of three roof styles (one is glass), four wheel styles, four seat colors, four dashboard prints, and three door-trim colors. There are three versions of the Nissan Leaf (S, SV, SL), each of which comes in seven colors (two whites, two grays, red, blue, black). How many different choices of car do you have?

For the Chevy Volt, there are $8 \cdot 3 \cdot 5 = 120$ ways to specify the car because we can have any combination of exterior, interior, and wheels. The Tesla has a ridiculous number of options: $9 \cdot 3 \cdot 4 \cdot 4 \cdot 4 \cdot 3 = 5184$.

In contrast, the $3 \cdot 7 = 21$ types of Nissan Leaf seem understated. Still, we add these three numbers together because we're only buying one car: $120 + 5184 + 21 = 5325$ choices of electric car.

5. Prove, or find a counterexample: the sum of two consecutive perfect cubes is odd.

Here are two proofs:

(1) First, observe that a number and its cube have the same parity.

Proof for even: An even number may be written as $2k$, and $(2k)^3 = 8k^3 = 2(4k^3) = 2q$, which is even.

Proof for odd: An odd number may be written as $2k+1$, and $(2k+1)^3 = 8k^3 + 12k^2 + 6k + 1 = 2(4k^3 + 6k^2 + 3k) + 1 = 2r + 1$, which is odd.

Now, consecutive perfect cubes have the property that one is odd and the other even. The sum of an odd number and an even number is odd ($2q + 2r + 1 = 2(q + r) + 1$), so the sum of any two perfect cubes where one is odd and the other even is odd.

(2) Either the consecutive perfect cubes are $(2k)^3$ and $(2k+1)^3$, or $(2k-1)^3$ and $(2k)^3$, depending on which of the numbers is even. In the first case, we have
$(2k)^3 + (2k+1)^3 = 8k^3 + 8k^3 + 12k^2 + 6k + 1 = 2(4k^3 + 4k^3 + 6k^2 + 3k) + 1$, which is odd. In the second case, we have
$(2k-1)^3 + (2k)^3 = 8k^3 - 12k^2 + 6k - 1 + 8k^3 = 2(4k^3 + 4k^3 - 6k^2 + 3k) - 1 = 2(4k^3 + 4k^3 - 6k^2 + 3k - 1) + 2 - 1 = 2(4k^3 + 4k^3 - 6k^2 + 3k - 1) + 1$, which is odd.

6. How many 4-digit phone extensions have no 0s and begin with 3?

   There is only one choice for the first digit (3) and 9 choices for each of the other three digits, so there are $1 \cdot 9 \cdot 9 \cdot 9 = 729$ such extensions.

7. In 2012, there were 3,952,841 live births in the US. (Source: http://www.cdc.gov/nchs/fastats/births.htm) Did there have to be two of these births within the same second?

   2012 was a leap year, so it had 366 days. Each of those days had 24 hours, each of which had 60 minutes, each of which had 60 seconds. Thus there were $366 \cdot 24 \cdot 60 \cdot 60 = 31,622,400$ seconds in 2012. There are more seconds than live births, so each birth could happen in a different second.

   However, there were only 527,040 minutes, so by the pigeonhole principle there must have been at least two births within the same minute—and $3,952,841/527,040 \approx 7.5$, so by the extended pigeonhole principle, there must have been at least 8 births that occurred during the same minute.

8. How many length-8 binary strings have no 0s in the fourth place?

   2^7, because the fourth place must be a 1 and there are two choices for each of the other 7 places.

9. You receive a choose-your-own-adventure certificate for a jewelry store! The deal is that you get to pick one of eight precious gems, and either a ring or a bracelet to put it in. There are three possible ring styles, and six possible bracelet styles.

   (a) How many possible prizes are there?

   (b) How did you answer the previous question? If you used the product principle first, re-answer the question using the sum principle first. (And if you used the sum principle first, re-answer the problem using the product principle first.)

   (c) On closer look, you realize that neither the ruby nor the emerald would look good on the bracelet. How many prizes are still possible?

   (a) 72.
   (b) Product principle then sum principle: $8 \cdot 3 + 8 \cdot 6$.
   Sum principle then product principle: $8 \cdot (3 + 6)$.

(c) Here, you have to use the product principle first: $8 \cdot 3 + 6 \cdot 6 = 60$.

10. I have a lot of stuff in my stuff-holder: Six ball-point pens, a silver star wand, three teal signature pens, a bronze-yellow colored pencil, five liquid ink pens, three mechanical pencils, a highlighter, six permanent markers, seven gel pens, a Hello Kitty lollipop, two markers, three wooden pencils, a 3-inch-long pen, a calligraphy marker, a pen shaped like a cat, and a pair of left-handed office scissors.

    How many writing utensils do I have in the stuff-holder?

    Good grief, that seems like a lot. But I just need to add up the numbers of things that are writing utensils: $6 + 3 + 1 + 5 + 3 + 1 + 6 + 7 + 2 + 3 + 1 + 1 + 1 = 40$. Seriously?

Chapter 2 🐦🐦

Sets and Logic

What's covered here? With sets, we have descriptions and constructions and notation. Within logic, we have truth tables and ways to describe reasoning. This is mostly pretty boring, and that's unavoidable. Still, you have my sympathies—you'll probably want to review the definitions in this chapter regularly.

2.1 Chapter 2 Definitions and Notation

2.1.1 Notation

$a_1, a_2 \in A$: Both a_1 and a_2 are elements of A.

$A \subset B$: A is a subset of B.

$\emptyset$: The empty set.

$\mathscr{P}(A)$: The power set of a set A.

$\overline{A}$: The complement of a set A.

$B \setminus A$: The complement of A relative to B.

$B - A$: Alternate notation for the complement of A relative to B.

$A \cup B$: The union of sets A and B.

$\bigcup_{i=1}^{n} A_i$: $A_1 \cup A_2 \cup \cdots \cup A_n$.

$A \cap B$: The intersection of sets A and B.

$\bigcap_{i=1}^{n} A_i$: $A_1 \cap A_2 \cap \cdots \cap A_n$.

$A \times B$: The Cartesian product of sets A and B.

$\wedge$: And.

$\vee$: Or.

$\neg$: Not.

$\Rightarrow$: Implies.

$\Leftrightarrow$: If and only if.

$\forall$: For all.

$\exists$: There exists.

2.1.2 Definitions

set: A mathematical object that contains distinct unordered elements. There may be finitely many or infinitely many elements in a set.

element: Elements can be words, objects, numbers, or sets (i.e., basically anything).

cardinality: The number of elements in a set.

subset: A is a subset of B if every element of A is also an element of B.

empty set: The set with no elements. Also called the null set.

null set: The empty set.

power set: The set of all subsets of A, denoted $\mathscr{P}(A)$.

set complement: If $A \subset B$, then $\overline{A} = B \setminus A$, all the elements of B that are not in A, is called the complement of A relative to B.

union: The union of sets A and B is a set $A \cup B$ containing all the elements in A and all the elements in B (with any duplicates removed). The union of many sets A_i contains all elements in the A_i (with any duplicates removed).

intersection: The intersection of sets A and B is a set $A \cap B$ containing every element that is in both A and B. The intersection of many sets A_i contains only elements that are in all of the A_i.

disjoint: Two sets A and B are called disjoint if $A \cap B = \emptyset$.

Cartesian product: The Cartesian product of sets A and B is a set $A \times B$ containing all possible ordered pairs where the first component is an element of A and the second component is an element of B. In other words, $A \times B = \{(a,b) \mid a \in A \text{ and } b \in B\}$. Likewise, the Cartesian product $A_1 \times A_2 \times \cdots \times A_n$ is the set of all n-tuples $(a_1, a_2, \ldots, a_n)$ where $a_i \in A_i$.

Venn diagram: A picture in which a big box denotes the universe of things under consideration and blobs represent sets. Venn diagrams are used to show relationships between sets.

statement: A sentence that is either true or false; it is the basic component of logical language. (To say that in a snooty way, a statement has a truth value from the set $\{true, false\}$.)

connective: A logical construction used to combine statements.

truth table: A table that lists the truth values of a statement.

and: The verbal analogue to set intersection, so P-and-Q is only true if both P and Q are true. Here is the corresponding truth table:

P	Q	$P \wedge Q$
T	T	T
T	F	F
F	T	F
F	F	F

or: The verbal analogue to set union, so P-or-Q is true whenever either P or Q is true. Here is the corresponding truth table:

P	Q	$P \vee Q$
T	T	T
T	F	T
F	T	T
F	F	F

xor: "Exclusive or." Here is the corresponding truth table:

P	Q	P xor Q
T	T	F
T	F	T
F	T	T
F	F	F

not: This gives a statement its opposite meaning; it makes a true statement false and makes a false statement true. Here is the corresponding truth table:

P	$\neg P$
T	F
F	T

P	Q	$P \Rightarrow Q$
T	T	T
T	F	F
F	T	T
F	F	T

implies: This means that one statement is a consequence of the other. Here is the corresponding truth table:

if-then: A statement involving implication.

conditional: An if-then statement.

P	Q	$P \Leftrightarrow Q$
T	T	T
T	F	F
F	T	F
F	F	T

if and only if: "P if and only if Q" is denoted $P \Leftrightarrow Q$ and means that the statements P and Q are logically equivalent. Here is the corresponding truth table:

iff : If and only if.

biconditional: An if-and-only-if statement.

quantifier: Quantifiers such as "for all" and "there exists" restrict the variables referred to in a statement.

implication: A statement of the form $P \Rightarrow Q$.

contrapositive: When $P \Rightarrow Q$ is the original statement, $\neg Q \Rightarrow \neg P$ is the contrapositive statement.

converse: When $P \Rightarrow Q$ is the original statement, $Q \Rightarrow P$ is the converse statement.

inverse statement: When $P \Rightarrow Q$ is the original statement, $\neg P \Rightarrow \neg Q$ is the inverse statement.

2.2   Chapter 2 Facts and Theorems

Fact.   If A is finite, then $|\mathscr{P}(A)| = 2^{|A|}$.

Facts about logical connectives and sets.

1. The logical *and* corresponds to set intersection, in the sense that the elements for which the statement $P \wedge Q$ holds are those in the set $A = \{x \mid P \text{ is true for } x\}$ *and* the set $B = \{x \mid Q \text{ is true for } x\}$, and together those elements form the set $A \cap B$.

2. The logical *or* corresponds to set union, in the sense that the elements for which the statement $P \vee Q$ holds are those in the set $A = \{x \mid P \text{ is true for } x\}$ *or* the set $B = \{x \mid Q \text{ is true for } x\}$, and together those elements form the set $A \cup B$.

3. The logical *not* corresponds to set complementation, in the sense that elements for which $\neg P$ holds are those not in the set $A = \{x \mid P \text{ is true for } x\}$; for this to make sense, we must make reference to a universe set U so that the elements not in A are those in $\overline{A}$, the complement of A relative to U.

Vague truth. Basically, if you have the statement $\neg(\forall \text{ stuff})$, that converts to $\exists \neg(\text{stuff})$, and if you have the statement $\neg(\exists \text{ stuff})$, that converts to $\forall \neg(\text{stuff})$.

DeMorgan's laws (logic version).

$\quad$ 🐤 $(\neg P) \vee (\neg Q)$ is logically equivalent to $\neg(P \wedge Q)$.

$\quad$ 🐤 $(\neg P) \wedge (\neg Q)$ is logically equivalent to $\neg(P \vee Q)$.

DeMorgan's laws (set version).

$\quad$ 🐤 $\overline{A} \cup \overline{B} = \overline{A \cap B}$.

$\quad$ 🐤 $\overline{A} \cap \overline{B} = \overline{A \cup B}$.

Equivalent statements. The statement $P \Rightarrow Q$ is logically equivalent to the statement $\neg Q \Rightarrow \neg P$.

2.3 Chapter 2 Proof Techniques: Double-Inclusion, Biconditionals ($\Longleftrightarrow$s), Proving the Contrapositive, and Proof by Contradiction

Double-Inclusion. To show that $A = B$, show first that $A \subset B$ and then show that $B \subset A$.

How to prove that $A \subset B$:

- ❦ Let a be any element of A.

- ❦ (Reasoning, statements.)

- ❦ Therefore, $a \in B$, and so $A \subset B$.

Biconditionals ($\Longleftrightarrow$s). The cleanest way to prove an if-and-only-if statement is to

(a) write ($\Rightarrow$) to indicate you'll prove that P implies Q (and then do so), and then

(b) write ($\Leftarrow$) to indicate you'll prove that Q implies P (and then do so).

Be sure to start a new paragraph for each implication.

Template for proving the contrapositive:

1. State the theorem in the form *if (conditions) are true, then (conclusion) is true*.

2. Restate the theorem in the equivalent form *if ¬(conclusion) is true, then ¬(conditions) is true*.

3. On a scratch sheet, write *assume* or *suppose ¬(conclusion) is true*.

4. Take some notes on what it means for ¬(conclusion) to be true. See where they lead.

5. Attempt to argue in the direction of ¬(conditions) is true.

6. Repeat attempts until you are successful.

7. Write up the results on a clean sheet, as follows:

🐤 Theorem: (State theorem here.)

🐤 Proof: Suppose ¬(conclusion) is true.

🐤 (Explain your reasoning in a logically airtight manner, so that no reader could question your statements.)

🐤 Therefore, ¬(conditions) is true, so our original theorem holds and we are done.

Template for a proof by contradiction:

1. Restate the theorem in the form *if (conditions) are true, then (conclusion) is true.*

2. On a scratch sheet, write *suppose not.* Then write out (conditions) and the negation of (conclusion).

3. Try to simplify the statement of ¬(conclusion) and see what this might mean.

4. Attempt to derive a contradiction of some kind—to one or more of (conditions) or to a commonly known mathematical truth.

5. Repeat attempts until you are successful.

6. Write up the results on a clean sheet, as follows:

   🐤 Theorem: (state theorem here)

   🐤 Proof: Suppose not. That is, suppose (conditions) are true but (conclusion) is false.

   🐤 (Translate this to a simpler statement if applicable. Derive a contradiction.)

   🐤 Contradiction!

   🐤 Therefore, (conclusion) is true. (Draw a box or checkmark or write Q.E.D. to indicate that you're done.)

2.4 Some Straightforward Examples of Chapter 2 Ideas

An example of manipulating set notation. Let $S_1 = \{q + 1 \in \mathbb{Z} \mid q = 2k$ for some $k \in \mathbb{Z}\}$ and let $S_2 = \{2r + 5 \mid r \in \mathbb{Z}\}$; we want to show that $S_1 = S_2$.

First, we will show that $S_1 \subset S_2$.

Let s be any element of S_1.

Then $s = 2k + 1$ for some $k \in \mathbb{Z}$.

If we let $r = k - 2$, then $s = 2k + 1 = 2(r + 2) + 1 = 2r + 5$, where $r \in \mathbb{Z}$, and therefore $s \in S_2$.

Now, we will show that $S_2 \subset S_1$.

Let t be any element of S_2.

Then $t = 2r + 5$, where $r \in \mathbb{Z}$.

Setting $k = r + 2$, we have that $t = 2r + 5 = 2(k - 2) + 5 = 2k + 1$, where $k \in \mathbb{Z}$, and therefore $t \in S_1$.

Because $S_1 \subset S_2$ and $S_2 \subset S_1$, we conclude that $S_1 = S_2$.

An example of Venn diagrams.   We will exhibit $(A \cap \overline{B}) \cup (\overline{A} \cap B)$ using Venn Diagrams.

We begin by looking within the parentheses. The first set of parentheses contains $A \cap \overline{B}$. We start at left in Figure 2.1 by hatching A. Because we want $A \cap \overline{B}$, we use a different hatching for $\overline{B}$ and then combine these so that $A \cap \overline{B}$ is crosshatched.

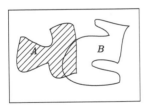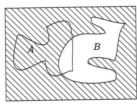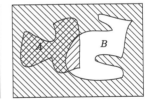

Figure 2.1. At left, A; in the middle, $\overline{B}$; at right, $A \cap \overline{B}$.

The second set of parentheses contains $\overline{A} \cap B$. We start at left in Figure 2.2 by hatching $\overline{A}$. Because we want $\overline{A} \cap B$, we use a different hatching for B and then combine these so that $\overline{A} \cap B$ is crosshatched.

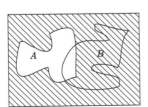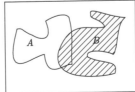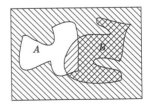

Figure 2.2. At left, $\overline{A}$; in the middle, B; at right, $\overline{A} \cap B$.

Finally, we combine these sets. We start at left in Figure 2.3 by showing $A \cap \overline{B}$, and in the middle we show $\overline{A} \cap B$. Because we want $(A \cap \overline{B}) \cup (\overline{A} \cap B)$, we display both at once using the same type of hatching.

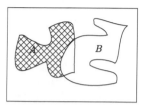 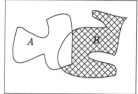 

Figure 2.3. At left, $A \cap \overline{B}$; in the middle, $\overline{A} \cap B$; at right, $(A \cap \overline{B}) \cup (\overline{A} \cap B)$.

An example of breaking down a very compound statement. Consider the statement

if $x \in \mathbb{Z}$ and $x > -7.2$ then x is positive or
$x \in \{0, -1, -2, -3, -4, -5, -6, -7\}$.

The largest logical substructure is the if-then implication, which combines the substatements

$x \in \mathbb{Z}$ *and* $x > -7.2$ and
x is positive or $x \in \{0, -1, -2, -3, -4, -5, -6, -7\}$.

Each of those has two substatements of its own; the *and* has substatements

$x \in \mathbb{Z}$ and
$x > -7.2$,

and the *or* has substatements

x is positive   and
$x \in \{0, -1, -2, -3, -4, -5, -6, -7\}$.

An example of order mattering. Let $d, e \in \mathbb{Z}$. Consider the statement $\forall e, \exists d$ such that $d < e$. This true statement basically says that given an integer, we can find a smaller one. For example, given $e = -32$, we can find $d = -4389$.

If we change the order of the quantifiers, our new statement is $\exists d, \forall e$ such that $d < e$. This statement says there is some integer such that every other integer is larger. That's not true!

(If you are (or have been) a student of calculus, compare this example to the formal (ε-δ) definition of limit.)

An example of wacky negations.   Consider the statement
for all futons, there exists a duck such that stripes are in fashion.
In logic notation, this becomes
$\forall$ *futons,* $\exists$ *a duck such that stripes are in fashion.* Thus, the negation
proceeds as
$\neg(\forall$ *futons,* $\exists$ *a duck such that stripes are in fashion*)
$\exists$ *a futon,* $\neg(\exists$ *a duck such that stripes are in fashion*)
$\exists$ *a futon, such that* $\forall$ *ducks* $\neg($*stripes are in fashion*) ... and finally,
there exists a futon such that for all ducks, stripes are not in fashion.

2.5   More Problems for Chapter 2

Those solutions that model a formal write-up (such as one might hand in
for homework) are to Problems 7 and 9.

1. On an October 2014 visit to the CVS Minute Clinic, the check-in
   kiosk asked the question, "If you have a copay for today's visit,
   will you be paying for it with a credit or debit card?"

   (a) Identify the formal logic quantifiers and structure in this question.

   (b) The visit in question was for a flu vaccine, which does not
   require a copay. The kiosk gave options of *Yes* and *No*. How
   should the visitor have answered?

   (c) Can you find a simpler way to word the question clearly? (In
   other words, what *should* the kiosk question ask?)

2. There was a recent campaign slogan heard on the radio: *Not just
   Blue Cross Blue Shield of Massachusetts, but Blue Cross Blue
   Shield ... of you.* Why is this mathematically nonsensical for res-
   idents of Massachusetts?

3. Consider the Venn diagram in Figure 2.4.

   (a) Express the shaded area as a set using unions, intersections,
   and/or complements of the sets Q, R, and S.

   (b) Let $Q = \{k \in \mathbb{Z} \mid |k| \leq 10\}$, $R =$ even numbers, and $S = \{n \in \mathbb{N} \mid n$ is a perfect square$\}$. List the elements of the shaded area.

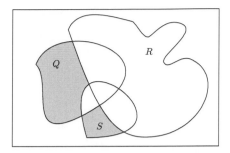

Figure 2.4. A Venn diagram of mystery.

4. Let $A = $ multiples of 4, and $B = $ multiples of 6. Write $A \cap B$ as a set in the form { sets | conditions }.

5. Negate the statement $\forall\, n \in \mathbb{Z}, \exists\, y \in 2\mathbb{N}$ *such that* $n = y \cdot k$ *for some* $k \in \mathbb{Z}$. Is either the statement or its negation true?

6. Prove that $k \in \mathbb{Z}$ is positive if and only if k^3 is positive.

7. Make a truth table for $\neg(P \wedge Q) \wedge ((P \vee Q) \wedge R)$. Can you express this statement (henceforth referred to as *aaaaaa!*) more simply?

8. Let $A = \{0, 1, 2\}$ and $B = \{1, 3, 5, 7\}$.

    (a) List the elements of $(A \times B) \cap (B \times A)$.

    (b) List the elements of $(A \setminus B) \times (B \setminus A)$.

9. Show that $(A \times B) \cup (C \times B) = (A \cup C) \times B$.

10. Show that $\{2k \mid k \in \mathbb{N}\} \cup \{4k + 1 \mid k \in \mathbb{W}\} \cup \{4k + 3 \mid k \in \mathbb{W}\} = \mathbb{N}$.

2.6   More Solutions for Chapter 2

1. On an October 2014 visit to the CVS Minute Clinic, the check-in kiosk asked the question, "If you have a copay for today's visit, will you be paying for it with a credit or debit card?"

    (a) Identify the formal logic quantifiers and structure in this question.

    (b) The visit in question was for a flu vaccine, which does not require a copay. The kiosk gave options of *Yes* and *No*. How should the visitor have answered?

    (c) Can you find a simpler way to word the question clearly? (In other words, what *should* the kiosk question ask?)

    (a) If $\exists$ (copay for today's procedure), then is the statement *I will pay with a credit or debit card* true? So, we have an $\exists$ and a $\Rightarrow$.

    (b) Because we have an implication of the form $P \Rightarrow Q$ with P false, we know from the truth table that $P \Rightarrow Q$ is true. But this doesn't say anything about whether we should answer *Yes* or *No*, because the question is about payment methods, not the truth of the statement. It shouldn't matter which answer is selected.

    (c) A better question might be, "If there are charges for today's visit, will you be paying for them with a credit or debit card?" That way it's not dependent on what is already known (there is no copay), but what might happen (additional services rendered, for example).

2. There was a recent campaign slogan heard on the radio: *Not just Blue Cross Blue Shield of Massachusetts, but Blue Cross Blue Shield ... of you.* Why is this mathematically nonsensical for residents of Massachusetts?

    You are in Massachusetts, so $\{you\} \subset$ Massachusetts and therefore you were already included in the described set. The slogan implies that Massachusetts $\subsetneq \{you\} \cup$ Massachusetts.

3. Consider the Venn diagram in Figure 2.4.

    (a) Express the shaded area as a set using unions, intersections, and/or complements of the sets Q, R, and S.

    (b) Let $Q = \{k \in \mathbb{Z} \mid |k| \leq 10\}$, $R =$ even numbers, and $S = \{n \in \mathbb{N} \mid n \text{ is a perfect square}\}$. List the elements of the shaded area.

(a) $(Q \setminus (R \cup S)) \cup (S \setminus (Q \cup R))$ is one way of expressing that area.

(b) The elements are $\{-9, -7, -5, -3, -1, 3, 5, 7, 25, 49, 81, \dots\}$

4. Let $A = $ multiples of 4, and $B = $ multiples of 6. Write $A \cap B$ as a set in the form $\{\text{ sets } \mid \text{ conditions }\}$.

$\{k \in \mathbb{Z} \mid k \text{ is a multiple of } 12\}$.

5. Negate the statement $\forall\, n \in \mathbb{Z}, \exists\, y \in 2\mathbb{N}$ *such that* $n = y \cdot k$ *for some* $k \in \mathbb{Z}$. Is either the statement or its negation true?

$\neg(\forall\, n \in \mathbb{Z}, \exists\, y \in 2\mathbb{N} \text{ such that } n = y \cdot k \text{ for some } k \in \mathbb{Z})$.
$\exists\, n \in \mathbb{Z}, \neg(\exists\, y \in 2\mathbb{N} \text{ such that } n = y \cdot k \text{ for some } k \in \mathbb{Z})$.
$\exists\, n \in \mathbb{Z}, \text{ such that } \forall\, y \in 2\mathbb{N}, \neg(n = y \cdot k \text{ for some } k \in \mathbb{Z})$.
$\exists\, n \in \mathbb{Z}, \text{ such that } \forall\, y \in 2\mathbb{N}, \text{ there is no } k \in \mathbb{Z} \text{ such that } n = y \cdot k$.

The statement is false; a counterexample is 1. The negation is true; consider any odd number.

6. Prove that $k \in \mathbb{Z}$ is positive if and only if k^3 is positive.

This is a biconditional, so it has two parts:

($\Rightarrow$) Suppose $k \in \mathbb{Z}$ is positive. Then k^3 is positive because the product of positive numbers is positive.

($\Leftarrow$) The simplest way of doing this is a proof by contrapositive, i.e., to prove that *if k^3 is positive, then k is positive* we'll show that *if k is not positive, then k^3 is not positive*. (It's difficult to be convincing about cube roots in a discrete context.)
Suppose $k \in \mathbb{Z}$ is negative.
Then k^3 is negative because the product of an odd number of negative numbers is negative.
This neglects the non-positive case of $k = 0$, but of course $0^3 = 0$ is not positive, so we are done.

7. Make a truth table for $\neg(P \wedge Q) \wedge ((P \vee Q) \wedge R)$. Can you express this statement (henceforth referred to as *aaaaaa!*) more simply?

P	Q	$P \vee Q$	R	$(P \vee Q) \wedge R$	$P \wedge Q$	$\neg(P \wedge Q)$	aaaaaa!
T	T	T	T	T	T	F	F
T	F	T	T	T	F	T	T
F	T	T	T	T	F	T	T
F	F	F	T	F	F	T	F
T	T	T	F	F	T	F	F
T	F	T	F	F	F	T	F
F	T	T	F	F	F	T	F
F	F	F	F	F	F	T	F

Notice that when R is false, so is the *aaaaaa!* statement. When R is true, we have the pattern of P xor Q (and also the pattern of $\neg(P \Leftrightarrow Q)$), so *aaaaaa!* is equivalent to $(P \text{ xor } Q) \wedge R$ (and also equivalent to $\neg(P \Leftrightarrow Q) \wedge R$).

8. Let $A = \{0,1,2\}$ and $B = \{1,3,5,7\}$.

   (a) List the elements of $(A \times B) \cap (B \times A)$.

   (b) List the elements of $(A \setminus B) \times (B \setminus A)$.

(a) Any element that is in both $A \times B$ and $B \times A$ must have each component in A and in B. Because $A \cap B = \{1\}$, the only element that qualifies is $(1,1)$.

(b) $A \setminus B = \{0,2\}$; $B \setminus A = \{3,5,7\}$.
Thus $(A \setminus B) \times (B \setminus A) = \{(0,3),(0,5),(0,7),(2,3),(2,5),(2,7)\}$.

9. Show that $(A \times B) \cup (C \times B) = (A \cup C) \times B$.

We proceed by double-inclusion.

If $x \in (A \times B) \cup (C \times B)$, then there are two cases: $x \in A \times B$ or $x \in C \times B$. In each case, $x \in (A \cup C) \times B$. One direction is done.

If $x \in (A \cup C) \times B$, then $x = (x_1, x_2)$ and $x_1 \in A$ or $x_1 \in C$. If $x_1 \in A$, then $x \in A \times B$; if $x_1 \in C$, then $x \in C \times B$. Thus, $x \in (A \times B) \cup (C \times B)$. Both directions are done.

10. Show that $\{2k \mid k \in \mathbb{N}\} \cup \{4k+1 \mid k \in \mathbb{W}\} \cup \{4k+3 \mid k \in \mathbb{W}\} = \mathbb{N}$.

We proceed by double-inclusion.

First, we show that $\{2k \mid k \in \mathbb{N}\} \cup \{4k+1 \mid k \in \mathbb{W}\} \cup \{4k+3 \mid k \in \mathbb{W}\} \subset \mathbb{N}$.

Consider $x \in \{2k \mid k \in \mathbb{N}\}$; because $k \in \mathbb{N}$, then $3k \in \mathbb{N}$ so $x \in \mathbb{N}$.
Now, let $y \in \{4k+1 \mid k \in \mathbb{W}\}$; for any $k \neq 0$, $k \in \mathbb{N}$ so $4k+1 \in \mathbb{N}$,

and if $k = 0$, then $4k + 1 = 1 \in \mathbb{N}$. Thus $y \in \mathbb{N}$. Similarly, we see that for $z \in \{4k+3 \mid k \in \mathbb{W}\}$, $z \in \mathbb{N}$.

Second, we show that $\{2k \mid k \in \mathbb{N}\} \cup \{4k+1 \mid k \in \mathbb{W}\} \cup \{4k + 3 \mid k \in \mathbb{W}\} \supset \mathbb{N}$.

Consider $n \in \mathbb{N}$. If n is even, then $n \in \{2k \mid k \in \mathbb{N}\}$. Otherwise, divide n by 4; the remainder must be 1 or 3 (as if it were 0 or 2, n would be even). If the remainder is 1, then $n \in \{4k + 1 \mid k \in \mathbb{W}\}$ and if the remainder is 3, then $\ell \in \{4k + 3 \mid k \in \mathbb{W}\}$. Therefore, $n \in \{2k \mid k \in \mathbb{N}\} \cup \{4k + 1 \mid k \in \mathbb{W}\} \cup \{4k + 3 \mid k \in \mathbb{W}\}$.

We conclude that $\{2k \mid k \in \mathbb{N}\} \cup \{4k + 1 \mid k \in \mathbb{W}\} \cup \{4k + 3 \mid k \in \mathbb{W}\} = \mathbb{N}$.

Chapter 3 🐤🐤🐤

Graphs and Functions

Graphs are the subject of roughly a third of the text, and here we have approximately a zillion related definitions as well as a formal way to tell when two drawings represent the same graph. This requires understanding of functions and their properties, so we actually start with that material.

3.1 Chapter 3 Definitions and Notation

3.1.1 Notation

$f : A \rightarrow B$: Usually a function, but sometimes a gipo.

$f|_D$: If $D \subset A$ and we want to talk about applying $f : A \rightarrow B$ just to elements of D, we write $f|_D$ to indicate that we are restricting the domain of f to D.

$\lfloor x \rfloor$: The floor function, which returns the integer equal to or just less than the input.

$\lceil x \rceil$: The ceiling function, which returns the integer equal to or just greater than the input.

1–1: One-to-one.

$V(G)$: The vertex set of a graph G.

$E(G)$: The edge set of a graph G.

$\{v_1, v_2\}$: An edge of a graph.

$v_1 v_2$: Also an edge of a graph.

P_n: A graph that is nothing but a path; it has n vertices and length $n - 1$.

C_n: A graph with n vertices that is nothing but a cycle.

W_n: A wheel graph with n vertices.

K_n: The complete graph with n vertices.

$K_{m,n}$: The complete bipartite graph with m vertices in one part and n vertices in the other part.

φ: A function; often, a proposed isomorphism.

$\star_A$: An operation on A, as in $a_1 \star_A a_2 = a_3$.

$A \cong B$: A is isomorphic to B.

$G \setminus e$ (or $G - e$): The graph G, but with the edge e removed and e's vertices left intact.

$G \setminus v$ (or $G - v$): The graph G but with the vertex v and all its incident edges removed.

$G \setminus H$ (or $G - H$): The graph G but with the subgraph H removed, and all edges incident to any vertex in H removed.

$\overline{G}$: Graph complement of G. Let G have n vertices; to form $\overline{G}$, we remove the edges of G from K_n, so that $\overline{G}$ has exactly the edges of K_n that G itself does not have.

$R(k,m)$: The Ramsey number $R(k,m)$ is the smallest number n such that a 2-edge-colored K_n must have either a K_k of one color or a K_m of the other color.

3.1.2 Definitions

function: We call $f : A \rightarrow B$ a function when, given any element a of the set A as input, the function f outputs a unique element $f(a) = b \in B$.

map: As a noun, map is a synonym for function; as a verb, it expresses the action a function takes, as in "f maps a to b."

well defined: The property that if $a_1 = a_2$, then $f(a_1) = f(a_2)$.

domain: The set A from which function inputs are taken.

target: The set B from which outputs are selected. Also called the target space.

image: The element $f(a)$ is called the image of the element a.

range: All the elements of the target space that are mapped to by the function; that is, for $f : A \rightarrow B$, the range of f is $Range(f) = \{f(a) \mid a \in A\}$.

gipo: A thing that *given* *input*, produces *output*, but is not necessarily well defined. (Notice that every function is a gipo, but not every gipo is a function.)

one-to-one: Whenever $f(a_1) = f(a_2)$, then $a_1 = a_2$. Every element of the target space is mapped to *at most* once. Also denoted 1–1.

injective: One-to-one.

into: One-to-one.

onto: For every $b \in B$, there exists some $a \in A$ such that $f(a) = b$. Every element of the target spacc is mapped to *at least* once.

surjective: Onto.

bijection: A function that is both one-to-one and onto. Every element of the target spacc is mapped to *exactly* once.

one-to-one correspondence: A bijection.

floor function: The floor function returns the integer equal to or just less than the input.

ceiling function: The ceiling function returns the integer equal to or just greater than the input.

graph: A pair $G = (V, E)$, where V is a set of dots and E is a set of pairs of vertices.

vertex: A dot, usually drawn as •, that can represent some object in a set of items.

vertices: Plural of *vertex*.

edge: A pair of vertices $e = \{v_1, v_2\}$ (sometimes abbreviated as $v_1 v_2$) that is usually represented by a line or curve between the dots representing v_1 and v_2.

adjacent: Two vertices joined by an edge are adjacent.

incident: An edge is incident to each of its endpoint vertices.

neighbor: Any vertex adjacent to a vertex v is a neighbor of v.

loop: An edge joining a vertex to itself.

multiple edge: More than one edge joining the same two vertices.

multiplicity: The number of edges in a multiple edge.

degree: The number of edges that emanate from a vertex.

degree sequence: A list of the degrees of the vertices in increasing
    order.

walk: A list of vertices alternating with edges, with both the start and end
    of the list vertices (not edges).

path: A walk where no vertices repeat.

cycle: A walk whose only repetition is the first/last vertex.

length: The number of edges of a path or cycle.

distance: The length of the shortest path between two vertices.

connected: A graph in which any two vertices are joined by some walk.

leaf: A vertex of degree 1.

tree: A connected graph with no cycles.

forest: A graph with no cycles.

simple graph: A graph that has no loops or multiple edges.

complete graph: A graph where every vertex is adjacent to every other
    vertex.

bipartite graph: A graph whose vertices can be separated into two piles,
    called parts, with edges between the parts, and no edges within
    either part.

complete bipartite graph: A bipartite graph with all possible edges; that
    is, if the parts are V_1, V_2, then every vertex in V_1 is adjacent to ev-
    ery vertex in V_2.

wheel graph: If you stick a vertex in the middle of an $(n-1)$-vertex cycle C_{n-1} (where $n-1$ is at least three) and connect it to all vertices on the cycle, you obtain the wheel graph, denoted W_n.

regular graph: A graph where all vertices have the same degree.

Petersen graph:

isomorphic graphs: Two graphs G, H are isomorphic if there exists a bijection $\varphi : V(G) \to V(H)$ such that $\{v_1, v_2\}$ is an edge in G if and only if $\{\varphi(v_1), \varphi(v_2)\}$ is an edge in H.

isomorphism: A gipo $\varphi : A \to B$ that is well-defined, one-to-one, onto, and preserves every operation defined on A (that is, if $\varphi(a_1 \star_A a_2) = \varphi(a_1) \star_B \varphi(a_2)$). In other words, an isomorphism is an operation-preserving bijection.

subgraph: A subgraph H of a graph G is a graph such that $V(H) \subset V(G)$ and $E(H) \subset E(G)$.

graph union: For graphs G_1 and G_2 (with disjoint vertex sets), the graph $G_1 \cup G_2$ is another graph G_3 (not connected) with $V(G_3) = V(G_1) \cup V(G_2)$ and $E(G_3) = E(G_1) \cup E(G_2)$.

graph component: An individual connected piece of a graph.

graph complement: Let G have n vertices; to form $\overline{G}$, we remove the edges of G from K_n, so that $\overline{G}$ has exactly the edges of K_n that G itself does not have.

weights: Numbers marked on the edges (or vertices) of a graph to indicate information such as distance or traffic capacity or population.

adjacency matrix: A matrix representing a graph, where each column and row corresponds to a vertex and each entry is the number of edges between the column vertex and row vertex.

Ramsey number: The Ramsey number $R(k, m)$ is the smallest number n such that a 2-edge-colored K_n must have either a K_k of one color or a K_m of the other color.

3.2   Chapter 3 Facts and Theorems

Nuances in the definition of *function*:

- 🐦 A function f on a domain A has to be defined on every single element of A. If some of them are skipped, either it's not a function after all or else it *is* a function, but secretly defined on some subset of A.

- 🐦 It is not cool to have two outputs for one input.

Fact 1.   If there is an injective function from A to B, then $|A| \leq |B|$.

Fact 2.   If there is a surjective function from A to B, then $|A| \geq |B|$.

Fact 3.   If there is a bijective function from A to B, then $|A| = |B|$.

Hey Hey Hey! Here's how sizes of sets say something about the pigeonhole principle.   Suppose $|A| > |B|$.
Then there is no injective function from A to B.
Therefore, every function from A to B must send at least two elements of A to a single element of B.
Let A represent pigeons and B represent pigeonholes. It follows that any function of pigeons to holes must place at least two pigeons in some hole.

Theorem 3.2.8.   Let A, B be finite sets and let f be a function $f : A \to B$. If $|A| = |B|$, then f is one-to-one $\Longleftrightarrow$ f is onto.

Counting functions in Section 3.3.1.   The number of functions from an m-element set to a q-element set is q^m. (There are q choices for the image of each of the m domain elements, so a total of $q \cdot q \cdots q = q^m$ possible functions.)
The number of *one-to-one* functions from an m-element set to a q-element set is $q \cdot (q-1) \cdots (q-(m-1))$, unless $m > q$ in which case there are no injective functions. (There are q choices for the image of the first domain element, but only $q-1$ choices for the image of the second domain element, and so on.)

Lemma 3.5.1, the handshaking lemma.   Because each edge is incident to two vertices, the sum of the degrees of the vertices of a graph must be twice the number of edges (and thus a multiple of two (i.e., even)). In symbols,

$$\sum_{v \in V(G)} \deg(v) = 2|E(G)|.$$

A simple caution. Sometimes "graph" means "simple graph" and sometimes it doesn't. This depends on who is speaking/writing and on the situation, so be aware that you may have to figure out whether the presence (or absence) of loops and/or multiple edges makes any difference.

3.3 Some Straightforward Examples of Chapter 3 Ideas

An example of a bijection proof. Let $g : \mathbb{W} \times \mathbb{Z}_2 \to \mathbb{Z}$ be defined by

$$g((n,t)) = \begin{cases} -n - 1 & \text{when } t = 0 \\ n & \text{when } t = 1. \end{cases}$$

We will show that g is a bijection.

First, we must show that g is injective.

Suppose $g((n_1, t_1)) = g((n_2, t_2))$. Then we have one of the following four cases, depending on the values of t_1, t_2:

1. $n_1 = n_2$ (in which case we're done).

2. $-n_1 - 1 = -n_2 - 1$, so that $n_1 = n_2$ (in which case we're done).

3. $-n_1 - 1 = n_2$, which is a contradiction because either n_1 or n_2 must be negative and there are no negative numbers in $\mathbb{W}$; thus, this case can't happen.

4. $n_1 = -n_2 - 1$, which cannot happen for exactly the same reasons. We conclude that g is one-to-one.

Second, we must show that g is surjective.

Consider $z \in \mathbb{Z}$. If $z < 0$, then $-z \in \mathbb{N}$ so that $-z - 1 \in \mathbb{W}$ and $g((-z - 1, 0)) = -(-z - 1) - 1 = z$. If $z = 0$, then $g((0, 1)) = 0$. If $z > 0$, then $g((z, 1)) = z$. Thus g is onto.

An example of many common graphs. Figure 3.1 shows examples of many common graphs, including complete graphs, complete bipartite graphs, a non-complete bipartite graph, cycle graphs, path graphs, wheel graphs, and a tree. Each graph is labeled.

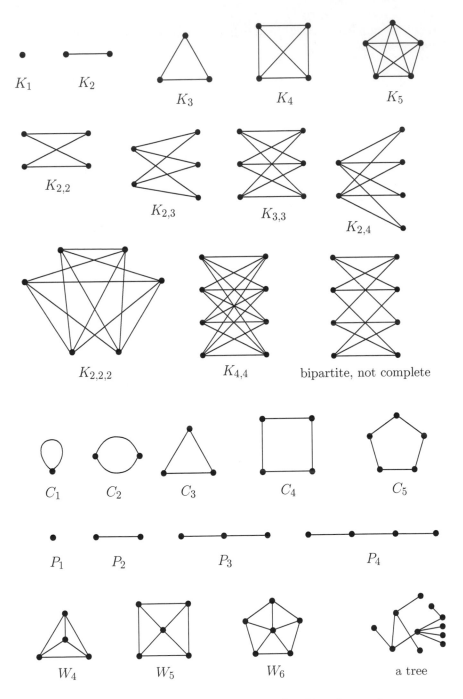

Figure 3.1. Some complete graphs, complete bipartite graphs, a non-complete bipartite graph, cycle graphs, path graphs, wheel graphs, and a tree.

3.4 More Problems for Chapter 3

Those solutions that model a formal write-up (such as one might hand in for homework) are to Problems 4, 6, and 9.

1. Let $S = \{s_1, s_2, \ldots, s_n\}$. How many functions are there with domain $\mathbb{Z}_3$ and target S? Of those functions, how many are one-to-one? How many are onto?

2. Draw all connected 3-regular graphs with four vertices.

3. Are the two graphs in Figure 3.2 isomorphic? Justify your response.

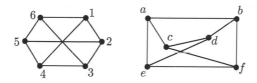

Figure 3.2. Two potentially isomorphic graphs.

4. Is the function $f : \mathbb{Z} \to \mathbb{Z}$ defined by $f(n) = \lfloor \sin(n) \rfloor$ a one-to-one function? Prove or disprove.

5. Is it possible to draw a graph with six vertices of degrees 2, 2, 3, 3, 4, and 4? If so, draw one. If not, explain why not.

6. A *finger-finger* graph is denoted by $F_{m,n}$ and has m fingers, from each of which grows n fingers; see Figure 3.3. Conjecture and

Figure 3.3. Some finger-finger graphs: $F_{1,4}, F_{2,2}, F_{3,5},$ and $F_{7,3}$.

prove formulae for the number of vertices and the number of edges of a finger-finger graph.

7. What can you say about the number of vertices of a 3-regular graph?

8. The following statement is true: *Any cycle C_n with $n \geq k$ has complement $\overline{C_n}$ containing a triangle.* Determine k and prove the statement.

9. Consider the map $g : (\mathbb{N} \times \mathbb{N}) \to \mathbb{N}$ defined by $g((a,b)) = ab$. Is this one-to-one? Onto? Give proofs.

10. Shown in Figure 3.4 are four infinite graphs in pairs A, B and C, D. One of these pairs is isomorphic and the other nonisomorphic. Which is which? Justify your response.

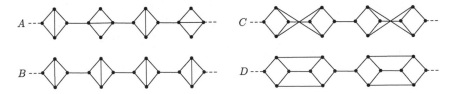

Figure 3.4. Who is who here?

3.5 More Solutions for Chapter 3

1. Let $S = \{s_1, s_2, \ldots, s_n\}$. How many functions are there with domain $\mathbb{Z}_3$ and target S? Of those functions, how many are one-to-one? How many are onto?

    There are n^3 functions from $\mathbb{Z}_3$ to S. Of those, $n \cdot (n-1) \cdot (n-2)$ are one-to-one.

    Only if $|S| \leq 3$ are any of the functions onto.

    If $|S| = 3$, then there are 6 onto functions (for the 6 different ways of assigning the three elements of $\mathbb{Z}_3$ to the three elements of S); if $|S| = 2$, then there are still 6 onto functions (two choices of which element of S is hit by only one element of $\mathbb{Z}_3$, and for each of those three choices of which element of $\mathbb{Z}_3$ goes there); if $|S| = 1$, then there is exactly one onto function (send all the elements of $\mathbb{Z}_3$ to the only element of S).

2. Draw all connected 3-regular graphs with four vertices.

    This is pretty much done by brute force, for example, by doing cases on the number of loops and the number of multiple edges in the graph. See Figure 3.5 for the 5 graphs.

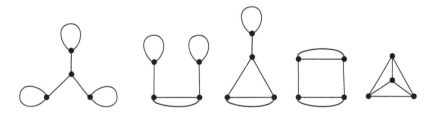

Figure 3.5. All 3-regular graphs with four vertices.

3. Are the two graphs in Figure 3.2 isomorphic? Justify your response.

    Yes, these are isomorphic. Use the map $\varphi(1) = a, \varphi(2) = b, \varphi(3) = f, \varphi(4) = c, \varphi(5) = d, \varphi(6) = e$. Many other isomorphisms are also possible.

4. Is the function $f : \mathbb{Z} \to \mathbb{Z}$ defined by $f(n) = \lfloor \sin(n) \rfloor$ a one-to-one function? Prove or disprove.

    The range of $\sin(x)$ is $[-1, 1]$. Thus, $\lfloor \sin(x) \rfloor$ can only take on the values $\{-1, 0, 1\}$. To show $f(n)$ is not one-to-one, we just have to find two values of n that both have $0 < \sin(n) < 1$. Trial and error gives $\sin(1) \approx 0.841471, \sin(2) \approx 0.909297$, so $f(1) = f(2)$; yet, $1 \neq 2$ and thus $f(n)$ is not one-to-one.

5. Is it possible to draw a graph with six vertices of degrees 2, 2, 3, 3, 4, and 4? If so, draw one. If not, explain why not.

    See Figure 3.6.

Figure 3.6. I'm a sixy, sixy graph.

6. A *finger-finger* graph is denoted by $F_{m,n}$ and has m fingers, from each of which grows n fingers; see Figure 3.3. Conjecture and prove formulae for the number of vertices and the number of edges of a finger-finger graph.

    A finger-finger graph $F_{m,n}$ has m edges for primary fingers, and from each primary finger n edges for secondary fingers. Thus by the sum and product principles, it has $m + mn$ edges.
    A finger-finger graph $F_{m,n}$ has one palm vertex, and one vertex from each of the m primary fingers, and then a vertex on the end of each secondary finger; there are n secondary fingers per primary finger, so by the sum and product principles, it has $1 + m + mn$ vertices.

7. What can you say about the number of vertices of a 3-regular graph?

    The number of vertices of a 3-regular graph must be even by the handshake lemma, as the total degree must be even and 3 is odd.

8. The following statement is true: *Any cycle C_n with $n \geq k$ has complement $\overline{C_n}$ containing a triangle.* Determine k and prove the statement.

    By trial and error, we find that $\overline{C_3}$ has no edges, $\overline{C_4}$ is two disjoint edges, and $\overline{C_5}$ is C_5, but $\overline{C_6}$ contains plenty of triangles. We suspect that $k = 6$. Now suppose $n \geq 6$, and number the vertices $1, 2, \ldots, n$. Vertices $1, 3, 5$ are not adjacent in C_n but form a triangle in $\overline{C_n}$.

9. Consider the map $g : (N \times N) \to N$ defined by $g((a,b)) = ab$. Is this one-to-one? Onto? Give proofs.

g is onto because given any $n \in \mathbb{N}$, the element $(1,n) \in \mathbb{N} \times \mathbb{N}$ is such that $g((1,n)) = 1 \cdot n = n$. However, g is not injective because we have $g((1,n)) = n = g((n,1))$ but $(1,n) \neq (n,1)$.

10. Shown in Figure 3.4 are four infinite graphs in pairs A, B and C, D. One of these pairs is isomorphic and the other nonisomorphic. Which is which? Justify your response.

Graphs A and B are nonisomorphic, because A has vertices of degree 2 but B does not.

Graphs C and D are isomorphic. Intuitively, we see that doing a vertical flip on one of each pair of diamonds will do the trick. We need to give a labeling to each graph and use this to define an isomorphism between the graphs. Figure 3.7 shows such a labeling.

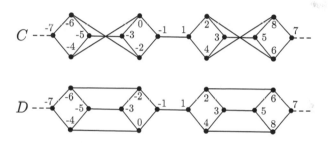

Figure 3.7. Oh, *that's* who you are.

Our isomorphism is defined by $\varphi(j) = j$, and this is clearly a bijection. To show formally that it preserves operations, we have to check that the adjacencies match. In both C and D, we can see the following:

* If $j = 8k + 1$ or $8k + 5$ $(k \in \mathbb{Z})$, j is adjacent to $j-1, j+1, j+3$.

* If $j = 8k + 2$ $(k \in \mathbb{Z})$, j is adjacent to $j-1, j+1, j+4$.

* If $j = 8k + 3$ or $8k + 7$ $(k \in \mathbb{Z})$, j is adjacent to $j-1, j+1, j+2$.

* If $j = 8k + 4$ $(k \in \mathbb{Z})$, j is adjacent to $j-3, j-1, j+4$.

* If $j = 8k + 6$ $(k \in \mathbb{Z})$, j is adjacent to $j-4, j-1, j+1$.

* If $j = 8k$ $(k \in \mathbb{Z})$, j is adjacent to $j-4, j-3, j-1$.

Chapter 4 🦆🦆🦆🦆

Induction

Yup. Induction. It's a proof technique related to recursion (the topic of a later chapter, Chapter 8).

4.1 Chapter 4 Definitions and Notation

4.1.1 Notation

$\sum_{j=m}^{n} f(j)$: The sum from $j = m$ to $j = n$ (that is, $n - m + 1$ terms) of an
expression written in terms of j. The summation sign $\sum$ indicates many additions.

$\sum_{j=m}^{n} f(j)$: Another way of writing $\sum_{j=m}^{n} f(j)$.

$n!$: Pronounced n-factorial, and shorthand for $n \cdot (n-1) \cdot (n-2) \cdot \ldots 3 \cdot 2 \cdot 1$.

4.2 Chapter 4 Facts and Theorems

DeMorgan's laws for n sets. Let $A_1, A_2, \ldots, A_n$ be n sets. Then for any $n \in \mathbb{N}$,

$$\overline{(A_1 \cap A_2 \cap \cdots \cap A_n)} = \overline{A_1} \cup \overline{A_2} \cup \cdots \cup \overline{A_n}$$

and

$$\overline{(A_1 \cup A_2 \cup \cdots \cup A_n)} = \overline{A_1} \cap \overline{A_2} \cap \cdots \cap \overline{A_n}.$$

4.3   Chapter 4 Proof Techniques: Induction (How Surprising!)

How to do a proof by induction:

- 🐦 **Base case.** Check to make sure that whatever you want to prove holds for small natural numbers, like 1, 2, or 3.

- 🐦 **Inductive hypothesis.** Assume that whatever you want to prove is true, as long as the variable in the statement is smaller than or equal to k; here, k is a specific (but unknown) value.

- 🐦 **Inductive step.** Consider the statement with $k+1$ as the variable. Use your knowledge that the statement is true when the variable is less than or equal to k in order to show that it's still true for $k+1$. (That is, use the base case(s) and inductive hypothesis.)

We might think of induction as a recipe for climbing a very branching ladder, in which case it goes like this:

(Base case) Make sure that we can get onto the ladder.

(Inductive hypothesis) Assume that if we happen to be on a k-level rung, we know how we got there.

(Inductive step) From a $(k+1)$-level rung, figure out how to get back to a k-level (or lower) rung, so we'll know where we are.

4.4   Some Straightforward Examples of Chapter 4 Ideas

An example of induction.   Let us show that if $n > 1$, then $3^n > 3n$.

(Base case) When $n = 2$, we see that $3^2 = 9$, and $9 > 6 = 3 \cdot 2$.

(Inductive hypothesis) For any $2 < n \leq k$, $3^n > 3n$.

(Inductive step) Consider the case $n = k+1$. We may rewrite $3^{k+1} = 3 \cdot 3^k$.

By the inductive hypothesis,

$3^k > 3k$, so $3 \cdot 3^k > 3 \cdot 3k$. Now,

$3 \cdot 3k = 9k = 3k + 6k = 3k + 3 + (6k - 3) > 3k + 3 = 3(k+1)$.

We know that $6k - 3 > 0$ because $k \geq 2$.

Combining the above statements, we have $3^{k+1} > 3(k+1)$ as desired.

Example 4.2.5 rewritten.   We will prove using induction that any tree with n vertices has $n - 1$ edges.

(Base case) We exhibit every tree with $n = 1, 2$ vertices in Figure 4.1.

Figure 4.1. The tree T_1 with one vertex (left) and the tree T_2 with two vertices (right).

Notice that T_1 has zero edges and that T_2 has one edge, so the base case holds.

(Inductive hypothesis) For any $n \leq k$, we assume that any tree with n vertices has $n - 1$ edges.

(Inductive step) Consider a generic tree, meaning any tree really, with $k + 1$ vertices; call it T. Choose any edge e of T.
Remove e from T to produce $T \setminus e$.
$T \setminus e$ is a pair of smaller trees S_1 and S_2; each has at least one vertex, and because together they have $k + 1$ vertices, each has fewer than $k + 1$ vertices.
Therefore, the inductive hypothesis applies to S_1 and to S_2.
Let the number of vertices of S_1 be r. Then S_2 has $k + 1 - r$ vertices. By the inductive hypothesis,
S_1 has $r - 1$ edges and S_2 has $k + 1 - r - 1$ edges.
$T = S_1 \cup e \cup S_2$, so by the sum principle T has $(r - 1) + 1 + (k + 1 - r - 1)$ $= r + k - r = k$ edges, as desired.

Example 4.4.1 rewritten. We will prove by induction that for $n \geq 2$, $2^n \leq 2^{n+1} - 2^{n-1} - 2$.

(Base case) When $n = 2$ we see that $2^2 = 4$ and $2^3 - 2^1 - 2 = 8 - 2 - 2 = 4$; this verifies the base case.

(Inductive hypothesis) For any $n \leq k$, $2^n \leq 2^{n+1} - 2^{n-1} - 2$.

(Inductive step) We would like to show that $2^{k+1} \leq 2^{k+2} - 2^k - 2$.
We know that $2^{k+1} = 2 \cdot 2^k$. The inductive hypothesis applies to 2^k, so
$2^{k+1} = 2 \cdot 2^k \leq 2 \cdot (2^{k+1} - 2^{k-1} - 2)$. Multiplying out gives the expression
$2^{k+1} \leq 2^{k+2} - 2^k - 4$. We know $-4 < -2$, so
$2^{k+1} \leq 2^{k+2} - 2^k - 4 < 2^{k+2} - 2^k - 2$ and we're done.

4.5   More Problems for Chapter 4

Seriously, do all of these problems by induction. That's what they're here for: induction practice. Those solutions that model a formal write-up (such as one might hand in for homework) are to Problems 2, 7, and 8.

1. Prove that $\sum_{j=1}^{n} 3 + 5j = \frac{1}{2}(11n + 5n^2)$.

2. Prove that $n^4 < 3 \cdot 8^n$.

3. Show that every convex polygon can be decomposed into triangles.

4. Show by induction that $K_{m,n}$ has mn edges.

5. Prove that $\sum_{j=0}^{n}(j+1)(j-2) = \frac{1}{3}(n-3)(n+1)(n+2)$.

6. Prove $(2(n!))^2 < 2^{(n!)^2}$ for sufficiently large values of n.

7. Use induction to prove the sum principle for n finite sets.

8. Take a piece of paper and fold it—not necessarily in half, but definitely with a single straight crease somewhere in the paper. Fold the (still folded) paper again. In fact, fold it n times, wherever you like. Now unfold it completely. Prove by induction that you can always color the paper with two colors (teal and purple) so that no fold line has the same color on both sides.

9. For what values of n is $5^{n+2} < 6^n$? Prove it.

10. Prove that any natural number ≥ 2 can be written as the product of prime numbers.

4.6   More Solutions for Chapter 4

1. Prove that $\sum_{j=1}^{n} 3 + 5j = \frac{1}{2}(11n + 5n^2)$.

   When $n = 1$, we have $3 + 5 = 8$ and $\frac{1}{2}(11 + 5) = \frac{16}{2} = 8$. Check.

   $\sum_{j=1}^{k+1} 3 + 5j = \sum_{j=1}^{k} 3 + 5j + 3 + 5(k+1)$. Using the inductive hypothesis,
   $= \frac{1}{2}(11k + 5k^2) + 3 + 5(k+1) = \frac{1}{2}(11k + 5k^2) + \frac{1}{2}(6 + 10(k+1))$
   $= \frac{1}{2}(11k + 5k^2 + 5 + 1 + 10k + 10) = \frac{1}{2}(11(k+1) + 5k^2 + 10k + 5) =$
   $\frac{1}{2}(11(k+1) + 5(k+1)^2)$ as desired.

2. Prove that $n^4 < 3 \cdot 8^n$.

   (Base case) When $n = 1$, we have $1^4 = 1 < 24 = 3 \cdot 8^1$.

   (Inductive hypothesis) For $n \leq k$, $k^4 < 3 \cdot 8^k$.

   (Inductive step) $(k+1)^4 = k^4 + 4k^3 + 6k^2 + 4k + 1$. The inductive hypothesis applies to k^4, so we have
   $(k+1)^4 < 3 \cdot 8^k + 4k^3 + 6k^2 + 4k + 1$.

   Now, we want to show that
   $3 \cdot 8^k + 4k^3 + 6k^2 + 4k + 1 < 3 \cdot 8^{k+1} = 24 \cdot 8^k$. If we can show that
   $4k^3 + 6k^2 + 4k + 1 < 21 \cdot 8^k$, that will do the trick (because then
   we'll have $3 \cdot 8^k + 4k^3 + 6k^2 + 4k + 1 < 3 \cdot 8^k + 21 \cdot 8^k = 24 \cdot 8^k$).

   We do know that $1 \leq k$, so
   $4k^3 + 6k^2 + 4k + 1 \leq 4k^3 + 6k^2 + 4k + k$, and $5k < 5k^2$, so
   $4k^3 + 6k^2 + 4k + k < 4k^3 + 6k^2 + 5k^2$, and $11k^2 < 11k^3$, so
   $4k^3 + 6k^2 + 6k^2 < 4k^3 + 11k^3$, and by the first example in this chapter we know that $15k^3 < 15 \cdot (2^k)^3 = 15 \cdot 2^{3k} = 15 \cdot 8^k < 21 \cdot 8^k$.

   Therefore
   $4k^3 + 6k^2 + 4k + 1 < 21 \cdot 8^k$, which means (from above) that
   $(k+1)^4 < 3 \cdot 8^k + 4k^3 + 6k^2 + 4k + 1 < 3 \cdot 8^k + 21 \cdot 8^k = 3 \cdot 8^{k+1}$.
   And we're done.

3. Show that every convex polygon can be decomposed into triangles.

   (Base cases) A triangle is already a triangle. Adding an edge joining opposite corners of a quadrilateral shows that the quadrilateral is composed of two triangles.

   (Inductive hypothesis) Suppose any convex polygon with $n \leq k$ sides can be decomposed into triangles.

(Inductive step) Consider a convex polygon with $k+1$ sides. Pick a vertex and travel along the edges of the polygon; skip the next vertex, but pick the one after that. Join these two vertices with an edge. On one side of the edge is a triangle made by the edge and the skipped vertex. On the other side of the edge is a convex polygon with $k-1$ sides. Therefore, by the inductive hypothesis it can be decomposed into triangles. Together with the triangle made by the edge and the skipped vertex, we have a decomposition of our $(k+1)$-sided polygon into triangles.

4. Show by induction that $K_{m,n}$ has mn edges.

(Base cases) Who knows what the smallest case should be here? Let's do three for good measure ... $K_{1,1}$ has $1 = 1 \cdot 1$ edge; $K_{1,2}$ has $2 = 1 \cdot 2$ edges; and, $K_{2,2}$ has $4 = 2 \cdot 2$ edges.

(Inductive hypothesis) For $m \leq n$ and $n \leq k$, $K_{m,n}$ has mn edges.

(Inductive step) Consider $K_{m,k+1}$, where $m \leq k+1$. (We can do this because $K_{m,n} = K_{n,m}$.) If we remove one of the $k+1$ vertices, we are left with $K_{m,k}$.
If $m \leq k$, then the inductive hypothesis applies and we know $K_{m,k}$ has mk edges. Replacing the removed vertex, we also restore m edges for a total of $mk + m = m(k+1)$ edges as desired.
If $m = k+1$, then $K_{m,k} = K_{k,k+1}$ and we may remove one of the $k+1$ vertices; this leaves us with $K_{k,k}$ which, by previous argument, has k^2 edges. Replacing the most recently removed vertex shows that $K_{k,k+1}$ has $k^2 + k = k(k+1)$ edges and replacing the first-removed vertex shows that $K_{k+1,k+1}$ has $k(k+1) + (k+1) = (k+1)^2$ edges, as desired.

5. Prove that $\sum_{j=0}^{n}(j+1)(j-2) = \frac{1}{3}(n-3)(n+1)(n+2)$.

When $n = 0$, we have $1(-2) = -2$ and $\frac{1}{3}(-3)(1)(2) = -2$. Check.

$\sum_{j=0}^{k+1}(j+1)(j-2) = \sum_{j=0}^{k}(j+1)(j-2) + (k+1+1)(k+1-2)$.
Using the inductive hypothesis,
$= \frac{1}{3}(k-3)(k+1)(k+2) + (k+2)(k-1)$
$= \frac{1}{3}(k^3 - 7k - 6) + \frac{1}{3}(3k^2 + 3k - 6)$
$= \frac{1}{3}(k^3 + 3k^2 - 4k - 12) = \frac{1}{3}(k+3)(k+2)(k-2)$
$= \frac{1}{3}(k+1-3)(k+1+1)(k+1+2)$ as desired.

6. Prove $(2(n!))^2 < 2^{(n!)^2}$ for sufficiently large values of n.

What are these sufficiently large values? Let's do some experiments:

n	$(2(n)!)^2$	$2^{(n!)^2}$
1	4	2
2	16	16
3	144	68,719,476,736

Looks like the statement holds for $n \geq 3$, and $n = 3$ is a base case.

Suppose that for $3 \leq n \leq k$, $(2(k!))^2 < 2^{(k!)^2}$.
Consider $((2(k+1)!))^2$.
We'll rewrite this as $(k+1)^2(2(k!))^2$ so we can use the inductive hypothesis to say that
$(k+1)^2(2(k!))^2 < (k+1)^2 2^{(k!)^2}$.

Now, $2^{((k+1)!)^2} = 2^{(k+1)^2(k!)^2} = 2^{(k+1)^2}2^{(k!)^2}$, so it remains to show that $(k+1)^2$ is less than $2^{(k+1)^2} = 2^{k^2+2k+1} = 2^{k^2}2^{2k}2$.

Because $2 < k$, we know $k+1 < k+k = 2k$, so
$(k+1)^2 < (2k)^2 = 4k^2$. By the first example in this chapter, we know that $k < 2^k$, so
$4k^2 < 4 \cdot 2^k2^k = 4 \cdot 2^{2k}$ and certainly $2 < 2^{k^2}$, so we can conclude that $(k+1)^2 < 2^{k^2}2^{2k}2$. And we're done!

7. Use induction to prove the sum principle for n finite sets.

(Base case) If A_1 has a_1 elements, then it has $\ldots a_1$ elements. Okay, that doesn't feel like we're saying anything. If A_1, A_2 have a_1, a_2 elements and $A_1 \cap A_2 = \emptyset$, then $A_1 \cup A_2$ has $a_1 + a_2$ elements.

(Inductive hypothesis) Let the finite set A_i have a_i elements, and let sets A_i and A_j be disjoint. Then for $n \leq k$, $\bigcup_{i=1}^{n} A_i$ has $\sum_{i=1}^{n} a_i$ elements.

(Inductive step) Consider $\bigcup_{i=1}^{k+1} A_i$. We may rewrite this as $\left(\bigcup_{i=1}^{k} A_i\right) \cup A_{k+1}$. The inductive hypothesis applies to $\bigcup_{i=1}^{k} A_i$ so we know it has $\sum_{i=1}^{k} a_i$ elements. And, $\left(\bigcup_{i=1}^{k} A_i\right) \cup A_{k+1}$ is the union of two sets so the base case applies and we know it has $\left(\sum_{i=1}^{k} a_i\right) + a_{k+1}$ elements. That last expression is simply $\sum_{i=1}^{k+1} a_i$ as desired.

8. Take a piece of paper and fold it—not necessarily in half, but definitely with a single straight crease somewhere in the paper. Fold the (still folded) paper again. In fact, fold it n times, wherever you like. Now unfold it completely. Prove by induction that you can always color the paper with two colors (teal and purple) so that no fold line has the same color on both sides.

(Base case) The base case is $n = 1$ fold. There are two regions in the paper, and one can be colored teal and the other purple.

(Inductive hypothesis) Suppose that for any $n \leq k$, a piece of paper folded n times and unfolded can be colored teal and purple so that no piece of a fold line has the same color on both sides.

(Inductive step) Consider a piece of paper that has been folded $k+1$ times. Unfold it once, and mark the fold line so that when you unfold it all the way you know which folds belong to that fold line. Unfold the paper completely and mark the entire $(k+1)st$ fold line. Now, ignore that line's presence—what you have is a piece of paper that has been folded k times and unfolded. This can be colored teal and purple so that no piece of a fold line has the same color on both sides.

Examine the $(k+1)st$ fold line. It bisects some regions. Leave the portions of those regions to one side (the first side) of the fold line alone, and switch the colors of the portions of regions on the other side (the second side) of the fold line—and switch the colors of all the whole regions to the second side of the fold line.

We need to show that there is no piece of a fold line that has the same color on both sides. There are three kinds of fold line pieces: those on the first side of the $(k+1)st$ fold line, those on the second side of the $(k+1)st$ fold line, and those that are part of the $(k+1)st$ fold line. Those on the first side already had different colors; those on the second side had different colors, both of which switched, so they still have different colors; and those part of the $(k+1)st$ fold line had the same color but one has been switched so there are now different colors on the two sides. Thus, we've correctly colored the paper teal and purple.

9. For what values of n is $5^{n+2} < 6^n$? Prove it.

By trial and error we note that for $n = 17$, we have
$5^{19} = 19{,}073{,}486{,}328{,}125 > 16{,}926{,}659{,}444{,}736 = 6^{17}$, but for
$n = 18$ we have
$5^{20} = 95{,}367{,}431{,}640{,}625 < 101{,}559{,}956{,}668{,}416 = 6^{18}$. (We generally expect that larger bases will produce larger functions in the long run.)

So we will use a base case of $n = 18$ and suppose that for $18 \le n \le k$, $5^{k+2} < 6^k$.

Consider $5^{k+3} = 5 \cdot 5^{k+2}$. By inductive hypothesis, we have $5 \cdot 5^{k+2} < 5 \cdot 6^k$, but we also know $5 < 6$, so $5 \cdot 6^k < 6 \cdot 6^k$ and the result is that $5^{k+3} < 6^{k+1}$ as desired.

10. Prove that any natural number ≥ 2 can be written as the product of prime numbers.

The base case is clear; 2 is the product of 2 (itself), which is prime. Suppose that any $n \le k$ can be written as the product of prime numbers, and consider $k + 1$.
Either $k + 1$ is prime, or it is not. If $k + 1$ is prime, then it is the product of one prime (itself). If $k + 1$ is not prime, then it is the product of (at least) two smaller numbers, each of which can be written as a product of primes (by the inductive hypothesis). Thus, the product of those products of prime numbers is *also* a product of prime numbers, and we are done.

Chapter 5 🐥🐥🐥🐥🐥

Algorithms with Ciphers

In this chapter, we discuss algorithms (crucial for computer science) and modular arithmetic (part of number theory, which is often a course on its own). Our primary examples of algorithms are ciphers, namely the shift cipher, the atbash cipher, and the Vigenère cipher.

5.1 Chapter 5 Definitions and Notation

5.1.1 Notation

$a \equiv b \pmod{n}$: a is congruent to b modulo n; this means that when a is divided by n, it leaves the same remainder as when b is divided by n.

$n|(a-b)$: This means $(a-b) = kn$ for some $k \in \mathbb{Z}$.

$\sim$: A verb meaning "is equivalent to."

$[a]$: All the elements equivalent to a using some equivalence relation, i.e., $\{s \in S \mid s \sim a\}$.

$\mathbb{Z}_n$: The integers modulo n, which as a set is $\{0, 1, \ldots, n-1\}$.

5.1.2 Definitions

algorithm: A finite list of unambiguous instructions to be performed on one or several inputs; some instructions may refer to others.

terminate: An algorithm that produces an output and ends after executing a finite number of instructions.

correct algorithm: An algorithm that does what it should do.

implementation: How an algorithm is made into executable code.

conditional: A statement within an algorithm that places conditions on an instruction.

if-then-else: A conditional that usually takes the form "if (conditions), then (action set 1), else (action set 2)" and is read/understood as "If (conditions) are met, then do (action set 1); otherwise, if (conditions) are *not* met, then do (action set 2)."

until: A conditional that takes the form "do (action set) until (conditions)" or "until (conditions), (action set)" and is read/understood as "Do (action set) until (conditions) are met and then go to the next instruction."

while: A conditional that takes the form "do (action set) while (conditions)" or "while (conditions), (action set)" and is read/understood as "Do (action set) while (conditions) hold, and when (conditions) are no longer met, go to the next instruction."

loop: An instruction to perform some set of actions more than once. (The instructions repeat, forming a string of instructions into a loop of instructions.)

iteration: The process of repeating instructions.

efficiency: A function that converts information about an algorithm's inputs into a real number that reflects how much work an algorithm needs to do in order to achieve a given goal.

running time: The output of an efficiency function for a given algorithm, usually described as a function type (e.g., linear, polynomial, exponential).

existence proof: A proof that shows that something exists.

constructive proof: A proof that produces an example of a desired object.

divides: Short for "divides evenly."

congruent modulo n: Two integers a and b are congruent modulo n when $(a-b) = kn$ for some $k \in \mathbb{Z}$.

integers modulo n: The set of different remainders obtainable by dividing integers by n.

symmetric property: This holds if when $s_1 \sim s_2$, then $s_2 \sim s_1$ for all $s_1, s_2 \in S$.

reflexive property: This holds if $s \sim s$ for all $s \in S$.

transitive property: This holds if when $s_1 \sim s_2$ and $s_2 \sim s_3$, then $s_1 \sim s_3$ for all $s_1, s_2, s_3 \in S$.

equivalence relation: An operation $\sim$ defined on a set S that satisfies the symmetric property, the reflexive property, and the transitive property.

equivalence class: All the elements equivalent to a using some equivalence relation, i.e., $[a] = \{s \in S \mid s \sim a\}$.

partition: A set of subsets $A_1, A_2, \ldots, A_n$ of a set A such that $A_1 \cup A_2 \cup \cdots \cup A_n = A$ and $A_i \cap A_j = \emptyset$ for all $i \neq j$.

encryption: The process of taking messages and converting them to forms that are not directly readable.

decryption: The process of converting received text to readable messages.

substitution cipher: A cipher that encrypts via letter-by-letter substitutions.

plaintext: A readable message.

ciphertext: A message encrypted with a known cipher.

wacktext: A communication we cannot read.

shift cipher: A cipher that encrypts by shifting each number by some fixed amount.

Caesar cipher: A shift cipher that shifts by 1.

ROT13: A shift cipher that shifts by 13 (and therefore encryption and decryption both proceed by adding 13 (mod 26).

atbash cipher: A substitution cipher that switches a with z, b with y, and so on; to encrypt, compute $-(\text{letter} + 1)$ (mod 26), and to decrypt, compute $-(\text{wackletter} + 1)$ (mod 26).

key word: A set of letters that provides the information needed to decrypt a cipher.

Vigenère cipher: A cipher that shifts each letter by an amount deter-
   mined by alignment with a key word. In modern usage, a Vigenère
   uses the key word repeatedly to decrypt an entire message; orig-
   inally, Vigenère himself used just one copy of the key word and
   then used the plaintext (or ciphertext) as it was generated for sub-
   sequent decryption (or encryption).

5.2   Chapter 5 Facts and Theorems

Operating with $\equiv$.   Let $a, b, c \in \mathbb{Z}$ and $n \in \mathbb{N}$.

   🦆 If $a \equiv b \pmod{n}$, then $b \equiv a \pmod{n}$.

   🦆 If $a \equiv b \pmod{n}$ and $b \equiv c \pmod{n}$, then $a \equiv c \pmod{n}$.

   🦆 (Theorem 5.3.3) If $a \equiv b \pmod{n}$, then $ac \equiv bc \pmod{n}$.

   🦆 If $a \equiv b \pmod{n}$ and $c \equiv d \pmod{n}$, then $ac \equiv bd \pmod{n}$.

   🦆 If $a \equiv b \pmod{n}$, then $a^k \equiv b^k \pmod{n}$.

Two facts about equivalence relations.

   1. If a set S has an equivalence relation $\sim$, then the equivalence classes
      of $\sim$ partition S.

   2. Congruence modulo n is an equivalence relation.

A table for converting letters to numbers for encryption/decryption.

Letter	a	b	c	d	e	f	g	h	i	j	k	l	m
Number	0	1	2	3	4	5	6	7	8	9	10	11	12
Letter	n	o	p	q	r	s	t	u	v	w	x	y	z
Number	13	14	15	16	17	18	19	20	21	22	23	24	25

5.3 Some Straightforward Examples of Chapter 5 Ideas

An example of an algorithm including conditionals and iteration. Let us suppose that we have a large supply of Jelly Babies but a limited supply of aliens (*n* aliens, to be precise).

1. Let *alien* = 1.

2. Face the *alien*th alien.

3. Pick up a Jelly Baby. Say, "Would you like a Jelly Baby?" to the alien in front of you.

4. If the alien responds positively, then hand it the Jelly Baby; otherwise, if the alien responds negatively, then shrug and eat the Jelly Baby yourself; otherwise, if the alien is impassive, then shake your head and continue.

5. If *alien* = *n*, return to your companions. Otherwise, continue.

6. Replace *alien* with *alien* + 1.

7. Go to step 2.

This algorithm includes a nested conditional; step 4 has the form *if A, then B, else (if C, then D, else (if E, then F))*. Note that the innermost if-then does not need an "else" because the only possibility left at that point is impassivity. (Well, okay, there might be some trouble in deciding whether an alien's response is positive, negative, or impassive, but those really are the only options.)

This algorithm also includes iteration. Steps 1, 5, and 6 collectively form a loop, by giving an initial value (step 1), iterating (step 6), and giving a condition on which we exit the loop (step 5).

An example of modular arithmetic in action. $22 \equiv 4 \pmod 6$ because 22 has a remainder of 4 when divided by 6.
Likewise, $40 \equiv 4 \pmod 6$, and $40 \equiv 22 \pmod 6 \ldots$ and $40 \equiv 745{,}408 \pmod 6$ as well.
$36 \equiv 24 \pmod{12}$, but while $4 \cdot 9 \equiv 4 \cdot 6 \pmod{12}$, notice that $9 \not\equiv 6 \pmod{12}$. (For the reader interested in more modular arithmetic, note that $9 \equiv 6 \pmod 3 \ldots$ what might be going on there?)

Example 5.4.1 rewritten.

Encryption with the Caesar cipher:
Start with the message   *duck.*
In numbers, this is   3 20 2 10.
We shift each number by 1, obtaining $(3+1) (20+1) (2+1) (10+1).$
This evaluates to   4 21 3 11.
In letters, this is   *evdl.*

Decryption with the Caesar cipher:
Start with the message   *ifo.*
In numbers, this is   8 5 14.
We shift each number by -1, obtaining $(8-1) (5-1) (14-1).$
This evaluates to   7 4 13.
In letters, this is   *hen.*

More Caesar cipher encryption:
Start with the message   *zebra.*
In numbers, this is   25 4 1 17 0.
This then encrypts to   26 5 2 18 1.
No letter corresponds to the number 26   so we must shift modulo 26.
We compute $(25 + 1 \ (\text{mod } 26)) \ (4 + 1$
                                              $(\text{mod } 26)) \ (1 + 1 \ (\text{mod } 26))$
                                              $(17 + 1 \ (\text{mod } 26)) \ (0 + 1$
                                              $(\text{mod } 26)).$

This evaluates to   0 5 2 18 1.
In letters, this is   *afcsb.*

More Caesar cipher decryption:
Start with the message   *tobaaz.*
In numbers, this is   19 14 1 0 0 25.
We shift each number by -1 (mod 26),
obtaining $(19 - 1 \ (\text{mod } 26)) \ (14 - 1$
                                              $(\text{mod } 26)) \ (1 - 1 \ (\text{mod } 26)) \ (0 - 1$
                                              $(\text{mod } 26)) \ (0 - 1 \ (\text{mod } 26))$
                                              $(25 - 1 \ (\text{mod } 26)).$

This evaluates to $18 \ 13 \ 0 \ -1 \ -1 \ 24.$
Because $-1 \equiv 25 \ (\text{mod } 26)$, our
message becomes   18 13 0 25 25 24.
In letters, this is   *snazzy.*

5.4   More Problems for Chapter 5

Those solutions that model a formal write-up (such as one might hand in for homework) are to Problems 3 and 9.

1. Find the smallest nonnegative integer x that satisfies the equation $3(x+7) \equiv 4(9-x)+1 \pmod 5$.

2. Encrypt this message from a supportive shark using a shift-by-10 cipher: YOU ARE SUPER GREAT AND FACES ARE HIGH IN PROTEIN

3. Prove, using only the definition of congruence modulo n, that if $a \equiv b \pmod n$, then $a+c \equiv b+c \pmod n$.

4. While you are distraught over your latest discrete math exam, a passerby shoves a scrap of paper into your hand that reads *xvghdib-hvivozz 21*. You suspect that this could be a shift cipher. What does the message say?

5. Here is an algorithm:

    (a) Get a pot, a cover, a stove, and an egg.

    (b) Put the egg in the pot.

    (c) Fill the pot with enough water to cover the egg.

    (d) Turn a burner to high heat.

    (e) Set the pot on the burner.

    (f) Put on a hat.

    (g) Wait until the water boils.

    (h) Wait for 3 minutes.

    (i) Remove the pot from the heat and add a cover.

    (j) Wait for 10 minutes.

    (k) Crack the shell of the egg.

    (l) Drain the water, replace with cold water, and let stand for 3 minutes.

    (m) Put away the egg.

    What are the inputs? What are the outputs? Does the algorithm terminate? What does the algorithm do? Are there any problems with this algorithm?

6. Let $a \sim b$ exactly when ab^2 is even. Is $\sim$ an equivalence relation?

7. Write an algorithm that lists the first 10 negative multiples of 9.

8. Encrypt *the foam shark visor is intended only for children* using the Vigenère cipher, using the key word *pickles*.

9. Encrypt *purple is the new green* and decrypt *xovzm zoo gsv gsrmth*, both using the atbash cipher.

10. Decrypt *xx ut e kcyrp nvavximtsfl ixoegwwpbggn* using a Vigenère cipher and the key word *pemberley*. Is this an original or a modern Vigenère cipher?

5.5 More Solutions for Chapter 5

1. Find the smallest nonnegative integer x that satisfies the equation $3(x+7) \equiv 4(9-x) + 1 \pmod 5$.

   That simplifies to
   $3x + 21 \equiv 36 - 4x + 1 \pmod 5$ which is also
   $3x + 1 \equiv 2 - 4x \pmod 5$ which is also
   $7x \equiv 1 \pmod 5$. So the question becomes, what is the smallest positive multiple of 7 that is one more than a multiple of 5? The positive multiples of 7 are $7, 14, 21, \ldots$ and there we find that $21 = 7 \cdot 3 \equiv 1 \pmod 5$, so
   $x = 3$.

2. Encrypt this message from a supportive shark using a shift-by-10 cipher: YOU ARE SUPER GREAT AND FACES ARE HIGH IN PROTEIN

   First we convert to numbers: $24, 14, 20, 0, 17, 4, 18, 20, 15, 4, 17, 6, 17,$ $4, 0, 19, 0, 13, 3, 5, 0, 2, 4, 18, 0, 17, 4, 7, 8, 6, 7, 8, 13, 15, 17, 14,$ $19, 4, 8, 13$.

   Then we add 10 $\pmod{26}$: $8, 24, 4, 10, 1, 14, 2, 4, 25, 14, 1, 16,$ $1, 14, 10, 3, 10, 23, 13, 15, 10, 12, 14, 2, 10, 1, 14, 17, 18, 16, 17, 18, 23,$ $25, 1, 24, 3, 14, 18, 23$.

   Then we convert back to letters:
   iyekbocezobqbokdkxnpkmockborsqrsxzbydosx and we're done.

3. Prove, using only the definition of congruence modulo n, that if $a \equiv b \pmod n$, then $a + c \equiv b + c \pmod n$.

   We know from the definition that
   $a - b = kn$. Therefore
   $a - b + c - c = kn$ or
   $a + c - (b + c) = kn$ so
   $a + c \equiv b + c \pmod n$.

4. While you are distraught over your latest discrete math exam, a passerby shoves a scrap of paper into your hand that reads *xvghdib-hvivozz 21*. You suspect that this could be a shift cipher. What does the message say?

   First convert to numbers: $23, 21, 6, 7, 3, 8, 1, 7, 21, 8, 21, 14, 25, 25$.

Then subtract 21 (mod 26) in the hopes that this was the shift: 2,0,11,12,8,13,6,12,0,13,0,19,4,4.

In letters, this is *calmingmanatee* or *calming manatee*. How nice to be handed a calming manatee!

5. Here is an algorithm:

   (a) Get a pot, a cover, a stove, and an egg.

   (b) Put the egg in a pot.

   (c) Fill the pot with enough water to cover the egg.

   (d) Turn a burner to high heat.

   (e) Set the pot on the burner.

   (f) Put on a hat.

   (g) Wait until the water boils.

   (h) Wait for 3 minutes.

   (i) Remove the pot from the heat and add a cover.

   (j) Wait for 10 minutes.

   (k) Crack the shell of the egg.

   (l) Drain the water, replace with cold water, and let stand for 3 minutes.

   (m) Put away the egg.

   What are the inputs? What are the outputs? Does the algorithm terminate? What does the algorithm do? Are there any problems with this algorithm?

   Inputs: a pot, a stove, a hat, and an egg.
   Outputs: A boiled egg.
   Terminate: Yes.
   Action: The algorithm boils an egg.
   Problems: The stove burner doesn't get turned off. There is an excess hat. The need for a sink/water is not mentioned as an input.

6. Let $a \sim b$ exactly when ab^2 is even. Is $\sim$ an equivalence relation?

   Let's check:
   Is $a \sim a$? Not always. a is even if and only if $aa^2 = a^3$ is even.
   If $a \sim b$, then is $b \sim a$? Yes. If ab^2 is even, then so is ab and thus so is ab^2.

If $a \sim b$ and $b \sim c$, then is $a \sim c$? No. We know that ab^2 is even and so is bc^2. The question is whether ac^2 is even. Suppose that b is even but a, c are odd; for example, let $a = 3, b = 4, c = 5$. Then $3 \sim 4$ because $3 \cdot 16 = 48$ is even, and $4 \sim 5$ because $4 \cdot 25 = 100$ is even, but $3 \not\sim 5$ because $3 \cdot 25 = 75$ is odd.

Therefore, this $\sim$ is not an equivalence relation.

7. Write an algorithm that lists the first 10 negative multiples of 9.

   (a) Set $k = 1$.

   (b) Output $-9 \cdot k$.

   (c) If $k = 10$, stop. Otherwise, continue.

   (d) Replace k with $k + 1$.

   (e) Go to step 2.

8. Encrypt *the foam shark visor is intended only for children* using the Vigenère cipher, using the key word *pickles*.

   The message becomes $19, 7, 4, 5, 14, 0, 12, 18, 7, 0, 17, 10, 21,$ $8, 18, 14, 17, 8, 18, 8, 13, 19, 4, 13, 3, 4, 3, 14,$ $13, 11, 24, 5, 14, 17, 2, 7, 8, 11, 3, 17, 4, 13$. That's 42 letters.
   The key word becomes $15, 8, 2, 10, 11, 4, 18$. That's 7 letters.
   How convenient! There are exactly 6 repetitions of the key word across the message.
   We add and get $8, 15, 6, 15, 25, 4, 4, 7, 15, 2, 1, 21, 25, 0, 7, 22, 19, 18, 3,$ $12, 5, 8, 12, 15, 13, 15, 7, 6, 2, 19, 0, 15, 25, 21, 20, 22, 16, 13,$ $13, 2, 8, 5$ which comes out as
   ipg pzee hpcbv zahwt sd mfimpnph gcta pzv uwqnncif when translated to letters.

9. Encrypt *purple is the new green* and decrypt *xovzm zoo gsv gsrmth*, both using the atbash cipher.

   purple is the new green becomes
   $15, 20, 17, 15, 11, 4, 8, 18, 19, 7, 4, 13, 4, 22, 6, 17, 4, 4, 13$.
   We compute $-(k+1) \pmod{26}$ for each number k and the result is
   $10, 5, 8, 10, 14, 21, 17, 7, 6, 18, 21, 12, 21, 3, 19, 8, 21, 21, 12$.
   In letters, this is *kfikov rh gsv mvd tivvm*.

   xovzm zoo gsv gsrmth becomes
   $23, 14, 21, 25, 12, 25, 14, 14, 6, 18, 21, 6, 18, 17, 12, 19, 7$.
   We compute $-(k+1) \pmod{26}$ for each number k and the result is

2, 11, 4, 0, 13, 0, 11, 11, 19, 7, 4, 19, 7, 8, 13, 6, 18.
In letters, this is *clean all the things*.

10. Decrypt *xx ut e kcyrp nvavximtsfl ixoegwwpbggn* using a Vigenère
cipher and the key word *pemberley*. Is this an original or a modern
Vigenère cipher?

In numbers, *xx ut e kcyrp nvavximtsfl ixoegwwpbggn* is
23, 23, 20, 19, 4, 10, 2, 24, 17, 15, 13, 21, 0, 21, 23, 8, 12, 19,
18, 5, 11, 8, 23, 14, 4, 6, 22, 22, 15, 1, 6, 6, 13, and *pemberley* is
15, 4, 12, 1, 4, 17, 11, 4, 24.
That's 9 characters long, so we subtract this from the first 9 numbers of the message, and obtain
8, 19, 8, 18, 0, 19, 17, 20, 19. In letters, this is
it is a trut. (Hm. I wonder what a "trut" is?)
In order to determine whether this is an original or a modern Vigenère cipher, we look at the next 9 letters of the message, namely
15, 13, 21, 0, 21, 23, 8, 12, 19 and
(1) subtract 15, 4, 12, 1, 4, 17, 11, 4, 24 to get 0, 9, 9, 25, 17, 6, 23, 8, 21
which translates to *ajjzrgxiv* and
(2) subtract 8, 19, 8, 18, 0, 19, 17, 20, 19 to get 7, 20, 13, 8, 21, 4, 17,
18, 0 which translates to *huniversa*.
The second makes a lot more sense than the first, so we will suppose this is an original Vigenère cipher and proceed:
Take the next 9 numbers, 18, 5, 11, 8, 23, 14, 4, 6, 22 and
subtract 7, 20, 13, 8, 21, 4, 17, 18, 0 to get 11, 11, 24, 0, 2, 10, 13, 14, 22
which translates to *llyacknow*.
Only 6 numbers remain, 22, 15, 1, 6, 6, 13, so we subtract 11, 11, 24, 0,
2, 10 to get 11, 4, 3, 6, 4, 3 which translates to *ledged*.
Our final text is *it is a truth universally acknowledged* ..., the
opening line of a famous novel.

Part II

Theme: Combinatorics

Chapter 6 🐤🐤🐤🐤🐤🐤

Binomial Coefficients and Pascal's Triangle

This is the first of four chapters on combinatorics topics. Here we have choice numbers (also known as binomial coefficients), which appear both in Pascal's triangle and the binomial theorem, as well as arrangements (or permutations). Two counting techniques are introduced, namely overcounting and compensation, and combinatorial proof.

6.1   Chapter 6 Definitions and Notation

6.1.1   Notation

$\binom{n}{k}$: n choose k, the number of ways one can choose k things from a pile of n different things.

$n!$: n factorial, $n \cdot (n-1) \cdot (n-2) \cdot \ldots 3 \cdot 2 \cdot 1$. By convention, $0! = 1$.

6.1.2   Definitions

choice number: A number of the form $\binom{n}{k}$ (that is, n choose k).

sugar number: A wrapped maple sugar candy in the shape of a number, used to represent a particular one of many objects.

combinatorial proof: A characteristic approach to solving combinatorics problems by counting some quantity in two different ways. Combinatorial proof works by looking at a set from multiple perspectives and gaining new information from each perspective.

Pascal's triangle: A triangular array of numbers in which the kth entry in the nth row is $\binom{n}{k}$. The top row is a single "1" and this is counted as the 0th entry of the 0th row.

factorial: An operation on a natural number n, denoted $n!$, that returns $n \cdot (n-1) \cdot (n-2) \cdot \ldots 3 \cdot 2 \cdot 1$.

permutation: An ordering of items.

binomial: A polynomial with exactly two terms, such as $2x - y$.

binomial coefficient: A coefficient in a binomial expansion, such as the coefficient "2" in $w^2 + 2wr + r = (w + r)^2$.

bijective proof: A proof that shows the equality of the sizes of two sets by demonstrating a one-to-one correspondence between two descriptions of the sets.

6.2   Chapter 6 Facts and Theorems

Borific side note.   There is a one-to-one correspondence between a handful of k sugar numbers and the k-element subset of $\{1, \ldots, n\}$ with those k numbers in it. It follows that $\binom{n}{k}$ is also the number of k-element subsets of an n-element set.

Pascal's identity, the most basic of choice notation identities.

$$\binom{n}{k} = \binom{n-1}{k-1} + \binom{n-1}{k}.$$

This corresponds to the generation of Pascal's triangle by adding pairs of entries in the previous row.

Pascal's triangle.   Well, the start of Pascal's triangle ...

$$
\begin{array}{ccccccccccccc}
 & & & & & & 1 & & & & & & \\
 & & & & & 1 & & 1 & & & & & \\
 & & & & 1 & & 2 & & 1 & & & & \\
 & & & 1 & & 3 & & 3 & & 1 & & & \\
 & & 1 & & 4 & & 6 & & 4 & & 1 & & \\
 & 1 & & 5 & & 10 & & 10 & & 5 & & 1 & \\
1 & & 6 & & 15 & & 20 & & 15 & & 6 & & 1 \\
 & & & & & & \vdots & & & & & &
\end{array}
$$

$\binom{n}{k}$ is the kth number in the nth row of the triangular array.

Fact and justification. $\binom{n}{k} = \binom{n}{n-k}$ because there are the same number of ways of letting k cats out of a bag as there are of letting $n-k$ cats out of a bag containing n cats.

Fact and justification. The sum of the numbers across a row of Pascal's triangle is 2^n because each entry in the row counts the number of subsets of $\{1,2,\ldots,n\}$ of a particular size, and every subset-size is represented.

Fact. Just as a one-to-one correspondence is a bijection and implies that two sets have the same size, a k-to-one correspondence indicates that the size of the domain is k times the size of the target space.

Counting permutations. Suppose we have n items and we want to know how many ways there are to order them. There are n choices for which item appears first in the ordering, then $n-1$ items remaining to choose for the second spot in the ordering, then $n-2$ items remaining to choose for the third spot in the ordering, and so forth. By the product principle, we have a total of $n \cdot (n-1) \cdot (n-2) \cdot \cdots \cdot 3 \cdot 2 \cdot 1$ ways to order n items.

Theorem 6.7.1, The binomial theorem.

$$(x+y)^n = \sum_{k=0}^{n} \binom{n}{k} x^{n-k} y^k = \binom{n}{0} x^n + \binom{n}{1} x^{n-1} y + \cdots + \binom{n}{n-1} x y^{n-1} + \binom{n}{n} y^n.$$

A formula for computing binomial coefficients.

$$\binom{n}{k} = \frac{n!}{k!(n-k)!}.$$

6.3 Chapter 6 Proof Techniques: Overcounting Carefully and Combinatorial Proof

How to overcount carefully. Suppose there are a things, and each of the a things has b aspects.

By the product principle, there are $a \cdot b$ variants in total.

But there might be a different way to think about these $a \cdot b$ variants. Perhaps the number $a \cdot b$ can be written as $c \cdot d$, where this represents c things that each have d aspects. (This would give a d-to-one correspondence between the set of $a \cdot b$ variants and the set of c things.)

In trying to solve a problem, we would probably know the values of a, b, and d, and be seeking the value of c.

Combinatorial proof. Combinatorial proof often begins with an equation. The goal is to interpret each side of the equation in a different way. The challenge in combinatorial proof is figuring out a good way to interpret a mathematical expression! We attempt to decide what we are counting if one side of the equation is the number of ways of choosing *something* and then answer the question, "Why does the other side of the equation count the same thing (but in a different way)?"

6.4   Some Straightforward Examples of Chapter 6 Ideas

An example of careful overcounting.   Consider a polygon with p sides. We would like to know how many vertices the polygon has. Each side of the polygon touches two vertices, so there are a total of $2p$ vertices. However, each vertex touches two sides, and that means we have overcounted by a factor of two. Thus the polygon has $\frac{2p}{2} = p$ vertices.

Example 6.5.2 rewritten.   How many different ways can we place three 4s on a 9×9 grid so that no two of them share a row or a column? Figure 6.1 shows a possibility and an impossibility. There are 9^2 squares, so there are 9^2 ways to place the first 4.

For each of these placements, we have used a row and a column. Therefore, there are only eight possibilities for the row and for the column of the second 4, for a total of 8^2 ways to place the second 4.

Neither of those two columns nor rows may be reused, so there are only 7^2 ways of placing the third 4.

This gives a total of $(9 \cdot 8 \cdot 7)^2$ ways of placing the three 4s.

However, this overcounts because placing the first 4 at (row 1 and column 2), and the second 4 at (row 3, column 4), gives the same result as placing the first 4 at (row 3, column 4), and the second 4 at (row 1, column 2).

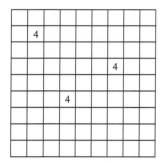 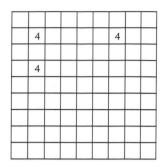

Figure 6.1. (Left) This may be a start on a solution. (Right) Well, *that's* not going to work ....

Every row on this board looks just like every other row, and the same is true for columns. We will exploit this symmetry.

For a given placement of three 4s, there are six ways we could have chosen it: let the placement be denoted by (a, b, c), where a (an ordered pair) represents the location of the first 4, b the location of the second 4, and c the location of the third 4. If we were to place the second 4 in location c and the third in location b, that'd be the same configuration of 4s; all possible permutations of (a, b, c) give this same result, and there are six of them. (The six permutations are $(a, b, c), (a, c, b), (b, a, c), (b, c, a), (c, a, b)$, and (c, b, a).)

So we have $\frac{(9 \cdot 8 \cdot 7)^2}{6} = \frac{254{,}016}{6} = 42{,}336$ ways of placing three 4s on a 9×9 grid.

Example 6.8.1 rewritten. Consider the identity $\sum_{k=0}^{n} \binom{n}{k} \binom{3n}{n-k} = \binom{4n}{n}$; we will produce a combinatorial proof of its validity. Using a story will be helpful.

We can interpret the right-hand side of the equation, $\binom{4n}{n}$, as the number of ways of choosing n teabags from a box of $4n$ teabags.

On the left-hand side of the equation appear choice notations with n and $3n$ in the top slots. These total to $4n$, so we will think of the teabags as n bags of peppermint tea and $3n$ bags of chamomile tea. Now $\binom{n}{k}$ is the number of ways of picking k of the n peppermint tea bags and $\binom{3n}{n-k}$ is the number of ways of choosing chamomile for the rest of our teabags.

However, the number of ways we can do that depends on what k is, and k could take any value from 0 to n. Therefore, we need to add up all the possibilities. This gives us $\sum_{k=0}^{n} \binom{n}{k} \binom{3n}{n-k}$ ways to choose n

teabags from $4n$ teabags, n of which are peppermint tea and $3n$ bags are chamomile tea.

We have now counted the same thing (number of ways of choosing n teabags from a box of $4n$ teabags) two different ways, so this completes our combinatorial proof.

6.5   More Problems for Chapter 6

Those solutions that model a formal write-up (such as one might hand in for homework) are to Problems 9 and 10.

1. Find a combinatorial proof for the identity $\displaystyle\sum_{k=0}^{n} k\binom{n}{k} = n2^{n-1}$.

2. Show that if n is even and k is odd, then $\binom{n}{k}$ is even.

3. Evaluate $\displaystyle\sum_{r=0}^{2m} 3^r 2^{2m-r}\binom{2m}{r}$.

4. The four students Ariel, Bingwen, Clarissa, and Dwayne have albums they need to listen to for a music appreciation class: *Duck Rock* (by Malcolm McLaren), *Duck Stab* (by The Residents), *Quack* (by Duck Sauce), and *This Time* (by Galapagos Duck).

    (a) How many ways are there to match the students with the albums?

    (b) The library has two listening rooms, each of which has two listening stations. How many ways are there to pair the students in the rooms?

    (c) Suppose the students have to sign up in advance, so they have to specify which listening station each student is using. Now how many ways are there for the students to be distributed into the rooms?

5. Give a combinatorial proof that $\binom{n}{4} = \frac{n!}{4!(n-4)!} = \frac{n(n-1)(n-2)(n-3)}{24}$.

6. At the art museum, you are decorating a round spinny top with stickers. However, this is an anti-creative art museum, so there are

only four equally spaced spots on the spinny top that are designated for receiving stickers, and there are only two colors of sticker
available—gray and grey. How many ways are there to "decorate"
the spinny top? (There are quotation marks because it is hard to
envision the spinny top as actually being decorated ...)

7. There are 18 students gathering to work on making a campus duck
   pond. They need to work in groups of three on various tasks. How
   many ways are there for the students to form groups?

8. Conjecture and prove a binomial identity for $\sum_{i=0}^{n} \binom{i}{5}$.

9. Find the coefficient of $x^4 y^6$ in $(5x^2 - 3y^3)^4$.

10. Prove that $\binom{2n}{2} = 2\binom{n}{2} + n^2$.

6.6   More Solutions for Chapter 6

1. Find a combinatorial proof for the identity $\sum_{k=0}^{n} k \binom{n}{k} = n2^{n-1}$.

   Let's say we need to make a store display of Pretty Rocks. One has to be "featured," meaning that it will be put on top of the Pretty Rocks sign. And we have a bin of n Pretty Rocks.

   The left-hand side of the equation says that we pick k of the n Pretty Rocks ($\binom{n}{k}$ ways) and then pick one of those to be featured ($\binom{k}{1} = k$ ways). But we didn't determine in advance how many Pretty Rocks we were going to put in the display, so we need to sum over k because we could have picked any number.

   The right-hand side of the equation says we pick a Pretty Rock to be featured (n choices for this) and then for each other Pretty Rock, we decide whether or not it's going to be in the display (2 choices for each of the $n-1$ remaining Pretty Rocks).

   This completes the proof.

2. Show that if n is even and k is odd, then $\binom{n}{k}$ is even.

   A direct proof might seem like a good idea here, but induction turns out to be *waaay* simpler. Compute a few base cases, and assume that the statement holds for "numerators" less than n. (We won't do this formally because the notation would be painful—k already means something here! But if we were being formal, we'd add a new letter to avoid confusion.)

   Note that $\binom{n}{k} = \binom{n-1}{k} + \binom{n-1}{k-1}$. This doesn't help (yet) because $n-1$ is odd and so the inductive hypothesis doesn't apply. So, we need to break it down further. That expression
   $= (\binom{n-2}{k} + \binom{n-2}{k-1}) + (\binom{n-2}{k-1} + \binom{n-2}{k-2}) = \binom{n-2}{k} + 2\binom{n-2}{k-1} + \binom{n-2}{k-2}$. That does it. The first and third terms are even because $n-2$ is even and the inductive hypothesis holds; the second term is clearly even because it's a multiple of 2. Done!

3. Evaluate $\sum_{r=0}^{2m} 3^r 2^{2m-r} \binom{2m}{r}$.

   This is the binomial theorem for $(3+2)^{2m} = 5^{2m}$.

4. The four students Ariel, Bingwen, Clarissa, and Dwayne have albums they need to listen to for a music appreciation class: *Duck*

Rock (by Malcolm McLaren), *Duck Stab* (by The Residents), *Quack* (by Duck Sauce), and *This Time* (by Galapagos Duck).

(a) How many ways are there to match the students with the albums?

(b) The library has two listening rooms, each of which has two listening stations. How many ways are there to pair the students in the rooms?

(c) Suppose the students have to sign up in advance, so they have to specify which listening station each student is using. Now how many ways are there for the students to be distributed into the rooms?

(a) $4!$.

(b) There are $\binom{4}{2}$ ways to put two students in one room, and then this determines who is in the other room.

(c) 6 ways to put the students in the rooms, and for each room there are two possible orderings of students to listening stations. So $6 \cdot 2 \cdot 2 = 24$.

OR, notice that this is just matching each student to one of four listening stations, so $4! = 24$.

5. Give a combinatorial proof that $\binom{n}{4} = \frac{n!}{4!(n-4)!} = \frac{n(n-1)(n-2)(n-3)}{24}$.

The left-hand side is the number of ways to choose 4 items from n items.

The right-hand side counts the number of ways to choose one item from n, then one item from the remaining $n-1$ items, then one item from the remaining $n-2$ items, and finally one item from the remaining $n-3$ items. If we don't care about the order in which these were picked, we need to divide by $4! = 24$.

But this is the same as the number of ways of just picking 4 things from n, so we're done.

6. At the art museum, you are decorating a round spinny top with stickers. However, this is an anti-creative art museum, so there are only four equally spaced spots on the spinny top that are designated for receiving stickers, and there are only two colors of sticker available—gray and grey. How many ways are there to "decorate" the spinny top? (There are quotation marks because it is hard to envision the spinny top as actually being decorated ...)

Let's see. For each sticker-spot, we could use grey (e) or gray (a), so there are $2^4 = 16$ possibilities. But the spinny top is round, so some of these are actually the same (e.g., *eaea* is the same as *aeae*). We suspect we can just account for our overcounting, but it's not straightforward. We have

all one color: *aaaa*, *eeee*.

one color different: *aeee* = *eaee* = *eeae* = *eeea*, and another four patterns all equivalent to *eaaa*.

two colors each: *eaea* = *aeae*, and *aaee* = *aeea* = *eeaa* = *eaae*.

That's all 16 accounted for, but only 6 are different.

So there are 6 ways to "decorate" the spinny top.

7. There are 18 students gathering to work on making a campus duck pond. They need to work in groups of three on various tasks. How many ways are there for the students to form groups?

We may place the first three students in a group in $\binom{18}{3}$ ways. Then we can choose the next group in $\binom{15}{3}$ ways, and so on for a total of $\binom{18}{3}\binom{15}{3}\binom{12}{3}\binom{9}{3}\binom{6}{3}\binom{3}{3}$ ways. However, the order in which we form the groups doesn't matter, so we need to divide by $6!$. By the way, that final answer is 190,590,400.

8. Conjecture and prove a binomial identity for $\displaystyle\sum_{i=0}^{n}\binom{i}{5}$.

We start by generating a few values—oops, this only makes sense for $n \geq 5$:

n	5	6	7	8	9
sum	1	7	28	84	210

Let's see. These are all numbers from Pascal's triangle, so what might they be? There are lots of 1s, but only a couple of 7s in Pascal's triangle—they're $\binom{7}{1}, \binom{7}{6}$. Looking nearby, we see $28 = \binom{8}{2} = \binom{8}{6}, 84 = \binom{9}{3} = \binom{9}{6}$, and $210 = \binom{10}{4} = \binom{10}{6}$. We conjecture that

$$\sum_{i=0}^{n}\binom{i}{5} = \binom{n+1}{6}.$$

Proof by induction: Base cases have already been checked, and we assume the statement holds for $n \leq k$. Here is the inductive step.

$$\sum_{i=0}^{k+1}\binom{i}{5} = \sum_{i=0}^{k}\binom{i}{5} + \binom{k+1}{5} = \binom{k+1}{6} + \binom{k+1}{5} = \binom{k+2}{6}$$

as desired.

9. Find the coefficient of $x^4 y^6$ in $(5x^2 - 3y^3)^4$.

First, let $a = 5x^2$ and let $b = -3y^3$.

From Pascal's triangle, we know the coefficients of $(a+b)^4$ are $(1,4,6,4,1)$.

The binomial theorem tells us we have $(5x^2)^4 + 4(5x^2)^3(-3y^3) + 6(5x^2)^2(-3y^3)^2 + 4(5x^2)(-3y^3)^3 + (-3y^3)^4$.

The monomial with variables x^4y^6 is $6(5x^2)^2(-3y^3)^2$, and it has coefficient $6 \cdot 5^2 \cdot (-3)^2 = 1350$.

10. Prove that $\binom{2n}{2} = 2\binom{n}{2} + n^2$.

We will do a combinatorial proof. $\binom{2n}{2}$ is the number of ways of choosing two books from a shelf with $2n$ books. On the other hand, we could split the shelf into two sub-shelves of n books each. We could choose two by picking one from each sub-shelf ($\binom{n}{1}^2 = n^2$), or by picking both from the same sub-shelf ($\binom{n}{2}$ for each sub-shelf).

Chapter 7 🐤🐤🐤🐤🐤🐤

Balls and Boxes and PIE: Counting Techniques

The main thrust of this chapter is figuring out how to deal with problems that can be phrased in terms of placing balls in boxes. This is harder than it sounds because

1. it's often difficult to figure out how to correctly phrase a problem and

2. not all types of balls-and-boxes problems have straightforward solutions.

Therefore, we only deal with labeled boxes. We also look at a few common techniques for solving counting problems that don't need balls-and-boxes formulations, namely slots and stars-and-bars. The Principle of Inclusion-Exclusion, commonly known as PIE, is an important side topic.

7.1 Chapter 7 Definitions and Notation

7.1.1 Notation

$\star$: A star, as used in the stars-and-bars strategy.

$|$: A bar, as used in the stars-and-bars strategy.

7.1.2 Definitions

order matters: "Item 1, then item 2" is different from "item 2, then item 1."

ordered: Order matters.

order doesn't matter: "Item 1, then item 2" is the same as "item 2, then item 1."

unordered: Order doesn't matter.

without repetition: Once an item has been used, it cannot be used again; for example, in forming a number with distinct digits.

repetition allowed: Once an item has been used, it can be used again; for example, in forming a number with possibly repeated digits.

stars-and-bars: A problem reframed so that it is about arranging stars and bars, and then solved.

inclusion-exclusion: A process of careful over- and undercounting in which we *include* individual set sizes by adding them (but this overcounts pairwise intersections) and *exclude* sizes of pairwise intersections by subtracting them (but this undercounts triple intersections) and include sizes of triple intersections by adding them ... and so on.

7.2   Chapter 7 Facts and Theorems

Balls-and-boxes problem types:

Question A   *How many ways are there to place k differently labeled balls, at most one per box, into n labeled boxes?* (order matters, without repetition)

Question B   *How many ways are there to place k identical (unlabeled) balls, at most one per box, into n labeled boxes?* (order doesn't matter, without repetition)

Question C   *How many ways are there to place balls, exactly one per box, with k different possible labels, into n labeled boxes?* (order matters)

Question D   *How many ways are there to place k unlabeled balls into n labeled boxes?* (order doesn't matter, repetition allowed)

Question D'   *How many ways are there to place k unlabeled balls into n labeled boxes, so that each box contains at least one ball?* (order doesn't matter, repetition allowed)

Question E   *How many ways are there to place k labeled balls into n labeled boxes, where k_j balls are placed into the jth box?*

Question F   *How many ways are there to rearrange the letters of a given word?*

How many ways . . . ... n labeled boxes?	at most one per box	any number per box	exactly one per box
k labeled (ordered) balls	**A:** $\binom{n}{k}k! = n(n-1)$ $\dots (n-k+1)$	**E, F:** (k_j balls unordered within box) $\frac{k!}{k_1!k_2!\dots k_n!} = \binom{k}{k_1}$ $\binom{k-k_1}{k_2}\binom{k-k_1-k_2}{k_3}$ $\dots \binom{k_n+k_{n-1}}{k_{n-1}}$	———
k unlabeled (unordered) balls	**B:** $\binom{n}{k}$	**D, D':** $\binom{k+n-1}{k} = \binom{k+n-1}{n-1}$ and $\binom{k-1}{n-1} = \binom{k-1}{k-n}$	———
unlimited balls, k different labels (order matters)	———	———	**C:** k^n

Table 7.1. Solutions summary.

Balls-and-boxes solutions summary. The grid in Table 7.1 presents our results in abbreviated form; each entry contains a question letter and the solution formula.

Strategy: slots. By *slot*, we mean an area reserved in which to place something, visualized as ___. Sometimes we can solve a problem by imagining filling some slots. For example, the number of possible two-character map locations is $26 \cdot 10$ because the first character must be a letter and the second must be a number.

Strategy: stars and bars. By *stars*, we mean $\star \star \star \dots \star$, and by *bars* we mean $| \, | \dots |$. Here is one possible arrangement of four stars and two bars: $\star \, | \, \star \star \, | \, \star$. When we reframe a problem so that it is about arranging stars and bars, we call the discussion a *stars-and-bars argument*.

If we are given s stars and b bars, then there are $\binom{s+b}{b}$ ways to arrange them, because there are $s+b$ spaces to fill with stars and bars and b of them must be chosen to be bars.

Suppose that instead we do not want any two bars to be adjacent. If we line up the stars, we see that there are $s - 1$ spots between the stars plus one spot to either side of the line, so a total of $s + 1$ spots in which bars can be placed. This gives us $\binom{s+1}{b}$ ways to arrange the stars and bars because we can put at most one bar in each spot.

Fact. Almost all of the counting done in this chapter (and the previous chapter) is accomplished via a series of bijections. Many of these are one-to-one correspondences between aspects of a problem and balls, boxes, slots, stars, or bars.

Principle of inclusion-exclusion. $|A \cup B| = |A| + |B| - |A \cap B|$ and $|A \cup B \cup C| = |A| + |B| + |C| - |A \cap B| - |A \cap C| - |B \cap C| + |A \cap B \cap C|$.

General PIE, Theorem 7.5.3. Let $A_1, A_2, \ldots, A_n$ be subsets of a finite set B. Then

$$\left| \bigcup_{i=1}^{n} A_i \right| = \sum_{j=1}^{n} (-1)^{j-1} \left(\sum_{\substack{\text{all } \binom{n}{j} \text{ intersections involving } j \text{ sets}}} \left| \bigcap A_i \right| \right).$$

7.3 Some Straightforward Examples of Chapter 7 Ideas

Example 7.2.2 rewritten. You have 12 young relatives and have knitted six different pairs of mittens. How many ways are there to give out the mittens at the wintertime family reunion? The children and pairs of mittens are all different, and no child should get more than one pair of mittens, so our question then becomes, "How many ways are there to place six labeled balls, at most one per box, into 12 labeled boxes?"

   The solution is that there are $\binom{12}{6}$ ways to choose the children to get mittens, and 6! ways to match the children to mittens, so there are $\binom{12}{6}6! = \frac{12!}{6!} = 665,280$ ways to distribute 6 different pairs of mittens to 12 young relatives.

Examples 7.2.5 and 7.4.7 rewritten. We have, for your example-ing pleasure, three sub-examples, two of which come in two variants.

   (1) Suppose an ogre is distributing 43 cupcakes to 12 baby mice. The baby mice are all different, and thus labeled. On the other hand, the cupcakes are all the same, and thus unlabeled. Therefore, the cupcakes correspond to balls and the baby mice correspond to boxes, so our question has become, "How many ways are there to place 43 unlabeled balls into 12 labeled boxes?" The answer is $\binom{43+12-1}{11} = 95,722,852,680$ ways to distribute 43 cupcakes to 12 baby mice.

We can also solve this by converting to stars and bars. Stars correspond to balls, and bars are the spaces between boxes. Now, our question has become, "How many ways are there to arrange 43 stars and 11 bars?" There are 54 slots in which a star or bar can be placed, so there are $\binom{54}{11} = \binom{54}{43}$ ways to arrange them.

Variant: Every baby mouse gets at least one cupcake. Therefore, we line up the baby mice and choose when each will stop munching and let the next one start. With c cupcakes, there are $c-1$ choices of times that a baby mouse could stop eating, and for b baby mice, there are $b-1$ transitions between baby mice eating. Therefore, the number of ways of distributing cupcakes is $\binom{c-1}{b-1}$.

(2) How many ways are there to line up five grey ducks and two white ducks? It's a bit of a stretch to make this into a balls-and-boxes problem—after all, nothing seems to be labeled here—but worth doing. Let us suppose that the grey ducks correspond to balls. The locations of the grey ducks (left of both white ducks, between the two white ducks, right of both white ducks) can be our labeled boxes. That transforms our question into, "How many ways are there to place five unlabeled balls into three labeled boxes?" The answer is that there are $\binom{5+3-1}{3-1} = \binom{7}{2} = 28$ ways. (We could instead let the white ducks correspond to balls, and obtain the question, "How many ways are there to place two unlabeled balls into six labeled boxes?" This has the same answer, $\binom{2+6-1}{2} = \binom{7}{2}$ ways.)

A simpler approach to this problem is to use slots: There are seven slots for ducks, and $\binom{7}{2}$ ways to place the white ducks; the grey ducks will occupy the remaining slots.

(3) How many nonnegative integer solutions are there to the equation $a+b+c = 5$? If we observe that $5 = 1+1+1+1+1$, we see that we need to distribute the 1s into a, b, c—so the 1s are unlabeled balls and the variables are labeled boxes. Now, our question has become, "How many ways are there to place five unlabeled balls into three labeled boxes?" The answer is $\binom{5+3-1}{3-1} = \binom{7}{2} = 28$.

Variant: How many *positive* integer solutions are there to the equation $a+b+c = 5$? We can put a 1 in each of the a, b, c variables, and then have two 1s left to distribute among the three variables. There are $\binom{2+3-1}{2} = 6$ ways to place two bars among three stars, and so six positive integer solutions to the equation $a+b+c = 5$.

Example 7.2.6 rewritten. At a party, there are 20 bite-sized pieces of fancily decorated strawberry cake and three fairies who decide that the first fairy will receive five, the second fairy will receive eight, and the third fairy will receive seven bites of strawberry cake. How many ways are there to distribute the fancily decorated strawberry cake bites among the fairies?

Each fairy gets multiple cake bites, so the fairies are boxes into which we put labeled cake-bite balls. Thus, our question is now, "How many ways are there to place 20 labeled balls into 3 labeled boxes, with those boxes receiving 5, 8, and 7 balls, respectively?"

The answer is that there are $\binom{20}{5}\binom{15}{8} = 99{,}768{,}240 = \frac{20!}{5!8!7!}$ ways to give 20 bites of fancily decorated strawberry cake to three fairies, so that the first of the fairies gets 5 bites, the second gets 8 bites, and the third gets 7 bites.

An example of apple PIE. A Cortland apple bushel has 35 apples, of which 20 are large, 25 are green, and 23 have stems. Ten of the apples are large, green, and have stems; 15 are large and green; 17 are green and have stems. How many of the Cortland apples are large and have stems?

The apple bushel has $|L \cup G \cup S| = 35$ apples, and we want to know $|L \cap S|$. We know from the PIE formula that
$$35 = (20 + 25 + 23) - (15 + 17 + |L \cap S|) + 10$$
$$= 68 - 32 - |L \cap S| + 10$$
$$= 46 - |L \cap S|, \text{ so}$$
$$|L \cap S| = 46 - 35 = 11.$$

7.4   More Problems for Chapter 7

Those solutions that model a formal write-up (such as one might hand in for homework) are to Problems 4 and 6.

1. Around Halloween, one can find bags of minipacks of SweeTarts. There are 3 SweeTarts in each pack, and the available color-flavors are orange, pink, purple, and blue.

   (a) How many different kinds of 3-SweeTart minipacks are there?

   (b) Actually, if you open a pack reasonably (instead of ripping it completely apart), you get only one SweeTart out to eat at a time. How many different experiences of 3-SweeTart minipacks are there?

2. In a 300-home neighborhood of Batamji, there are four different kinds of trees (magnolias, cypress, willow, and river birch). Forty homes have just cypress trees; 32 homes have just willow trees; 9 homes have just river birch. Seventy homes have magnolia and willow; 47 homes have magnolia and cypress; 40 homes have cypress and river birch; 61 homes have magnolia and river birch; 44 homes have cypress and willow; 56 homes have willow and river birch. Twelve homes have magnolias, cypress, willow, and river birch; 38 homes have magnolias, cypress, and willow; 19 homes have magnolias, willow, and river birch; 28 homes have magnolias, cypress, and river birch; 29 homes have cypress, willow, and river birch. How many homes have just magnolia trees?

3. A hungry ninja is making tacos with the following ingredients: beans, guacamole, cheese, tomatoes, scallions, salsa, and lettuce. How many ways can the ninja assemble tacos for different meals (breakfast, snack, lunch, tea, dinner), the first of which has 3 fillings, the next two of which have 4 fillings, and the final 2 of which have 5 fillings?

4. The Edgy Ruck company uses length-10 serial numbers that mix letters (except Y) and numbers. How many serial numbers are there that have a 7 in the fourth slot and a consonant in the eighth slot, or have a letter in the fifth slot and a vowel in the ninth slot?

5. All that is left of your Hello Kitty Jelly Belly sampler is the 12 Very Cherry flavored Jelly Bellies (because you *hate* that flavor) and you have 4 friends who volunteer to eat them for you. How many ways are there to hand out the Jelly Bellies?

6. You've made a pile of 8 cute notes for your best friend to find. She has 12 folders, one for each of her classes and activities. How many ways are there to tuck the notes into folders? (Of course you will not put more than one note in a folder. That would be excessive.)

7. The computer print-out says it all: Your first student needs 3 Learning Modules inserted, your second student needs 5 Learning Modules inserted, and your third student needs 54 Learning Modules inserted from the bank of 62 new government-approved-topic Learning Modules. But wait . . . The computer print-out doesn't say which Learning Modules should go to which student. How many ways can you assign Learning Module topics to students?

8. Your spiky little plant has once again outgrown its pot, and when you split off all the small bits into different pots you discover you have 23 spiky plant-spawn. You've promised 8 people they can have baby spiky plants, but really you want to get rid of *all* of the spiky plant-spawn so they don't take over your house. How many ways are there to distribute the 23 baby spiky plants to the 8 people?

9. How many anagrams are there of the word ENUMERATE?

10. How many ways are there to list the 50 U.S. states so that no two states beginning with "A" are next to each other?

7.5   More Solutions for Chapter 7

1. Around Halloween, one can find bags of minipacks of SweeTarts. There are 3 SweeTarts in each pack, and the available color-flavors are orange, pink, purple, and blue.

   (a) How many different kinds of 3-SweeTart minipacks are there?

   (b) Actually, if you open a pack reasonably (instead of ripping it completely apart), you get only one SweeTart out to eat at a time. How many different experiences of 3-SweeTart minipacks are there?

   (a) Here we have three balls—the SweeTarts—and four boxes—the color-flavors—in which to place them. A box can get any number of balls (including 0). The balls aren't labeled, so we have **Question D** which has solution $\binom{4+3-1}{3} = \binom{6}{3} = 20$. So 20 different kinds of minipacks.

   (b) 4^3 because there are four choices for the first, second, and third SweeTart color-flavors.

2. In a 300-home neighborhood of Batamji, there are four different kinds of trees (magnolias, cypress, willow, and river birch). Forty homes have just cypress trees; 32 homes have just willow trees; 9 homes have just river birch. Seventy homes have magnolia and willow; 47 homes have magnolia and cypress; 40 homes have cypress and river birch; 61 homes have magnolia and river birch; 44 homes have cypress and willow; 56 homes have willow and river birch. Twelve homes have magnolias, cypress, willow, and river birch; 38 homes have magnolias, cypress, and willow; 19 homes have magnolias, willow, and river birch; 28 homes have magnolias, cypress, and river birch; 29 homes have cypress, willow, and river birch. How many homes have just magnolia trees?

   Let's see here. This is a little bit tricky because we're given some numbers of the form $|A \setminus (B \cup C \cup D)|$ ("just cypress") and others that are $|E \cap F|$ ("willow and river birch") ... that means we only have to deal with overlaps starting with the paired intersections. We have $300 = M + 40 + 32 + 9 + [(70 + 47 + 40 + 61 + 44 + 56) - (38 + 19 + 28 + 29) + 12] = M + 297$, so there are 3 homes with only magnolia trees.

3. A hungry ninja is making tacos with the following ingredients: beans, guacamole, cheese, tomatoes, scallions, salsa, and lettuce. How many ways can the ninja assemble tacos for different meals (breakfast, snack, lunch, tea, dinner), the first of which has 3 fillings, the next two of which have 4 fillings, and the final 2 of which have 5 fillings?

    We have labeled ingredients (balls) and labeled tacos (boxes), so this looks like **Question E**. However, that assumes we don't have repetition of the 7 balls, and we need that for our 21 fillings! Here is how we proceed—we count the number of ways there are to construct each taco. For the first taco, we have $\binom{7}{3}$ toppings, for the second two we have $\binom{7}{4}$ toppings, and for the last two we have $\binom{7}{5}$ toppings, for a total of $\binom{7}{3} \cdot \binom{7}{4} \cdot \binom{7}{4} \cdot \binom{7}{5} \cdot \binom{7}{5} = 18,907,875$ meal possibilities for the ninja's day.

4. The Edgy Ruck company uses length-10 serial numbers that mix letters (except Y) and numbers. How many serial numbers are there that have a 7 in the fourth slot and a consonant in the eighth slot, or have a letter in the fifth slot and a vowel in the ninth slot?

    First, there are 25 letters, 5 of which are vowels and 20 of which are consonants, and 10 numbers. We will use slots and PIE for this problem.

    Serial numbers with a 7 in the fourth slot and a consonant in the eighth slot: $35 \cdot 35 \cdot 35 \cdot 1 \cdot 35 \cdot 35 \cdot 35 \cdot 20 \cdot 35 \cdot 35 = 45,037,507,812,500$.

    Serial numbers with a letter in the fifth slot and a vowel in the ninth slot: $35 \cdot 35 \cdot 35 \cdot 35 \cdot 25 \cdot 35 \cdot 35 \cdot 35 \cdot 5 \cdot 35 = 281,484,423,828,125$.

    Serial numbers with a 7 in the fourth slot and a letter in the fifth slot and a consonant in the eighth slot and a vowel in the ninth slot: $35 \cdot 35 \cdot 35 \cdot 1 \cdot 25 \cdot 35 \cdot 35 \cdot 20 \cdot 5 \cdot 35 = 4,595,664,062,500$.

    PIE says we have $45,037,507,812,500 + 281,484,423,828,125 - 4,595,664,062,500 = 321,926,267,578,125$ serial numbers with 7 in the fourth slot and a consonant in the eighth slot, or have a letter in the fifth slot and a vowel in the ninth slot.

5. All that is left of your Hello Kitty Jelly Belly sampler is the 12 Very Cherry flavored Jelly Bellies (because you *hate* that flavor) and you have 4 friends who volunteer to eat them for you. How many ways are there to hand out the Jelly Bellies?

This is **Question D** because you have a dozen unlabeled balls and 5 labeled boxes, so the answer is $\binom{12+4-1}{4} = \binom{16}{4} = 1820$ ways.

6. You've made a pile of 8 cute notes for your best friend to find. She has 12 folders, one for each of her classes and activities. How many ways are there to tuck the notes into folders? (Of course you will not put more than one note in a folder. That would be excessive.)

   The notes are all different and so are the folders, so we have labeled balls and labeled boxes. There are 12 choices for the folder for the first note, then 11 for the second, and so forth—$12 \cdot 11 \cdot 10 \cdot 9 \cdot 8 \cdot 7 \cdot 6 \cdot 5 = 19,958,400$ is the count.

7. The computer print-out says it all: Your first student needs 3 Learning Modules inserted, your second student needs 5 Learning Modules inserted, and your third student needs 54 Learning Modules inserted from the bank of 62 new government-approved-topic Learning Modules. But wait ... The computer print-out doesn't say which Learning Modules should go to which student. How many ways can you assign Learning Module topics to students?

   This is a classic **Question E** problem:
   $\binom{62}{3} \cdot \binom{62}{5} \cdot \binom{62}{54} = 827,467,389,801,458,843,800$. Hm. Guess the computer must not care much about which way you assign the topics.

8. Your spiky little plant has once again outgrown its pot, and when you split off all the small bits into different pots you discover you have 23 spiky plant-spawn. You've promised 8 people they can have baby spiky plants, but really you want to get rid of *all* of the spiky plant-spawn so they don't take over your house. How many ways are there to distribute the 23 baby spiky plants to the 8 people?

   Each person must get at least one plant, so we are in **Question D'** and so the answer is $\binom{23-1}{8-1} = \binom{22}{7} = 170,544$.

9. How many anagrams are there of the word ENUMERATE?

   ENUMERATE has 9 letters, three of which are Es. Thus there are $9!/3! = 60,480$ anagrams.

10. How many ways are there to list the 50 U.S. states so that no two states beginning with "A" are next to each other?

Oh, good grief. What *are* the states that start with "A"? Alaska, Arkansas, Arizona, and Alabama. So we have 46 states that can be ordered any-which-way, and then we need to stick the 4 "A" states into the 47 places between and around those 46 states.

There are 46! ways to order the 46 non-"A" states, and for each of those, we have $\binom{47}{4}$ ways to place the 4 "A" states. Except that for each of those placements, there are 4! ways we could have ordered the "A" states, so in total we have $46! \cdot \binom{47}{4} \cdot 4!$ which is a number with 65 digits, so I will not write it out. Okay, fine. It's 23,555,404,836,837,197,892,961,193,467,390,979,135,594,520, 358,001,049,600,000,000,000.

This problem can also be solved using PIE, but that solution is much more complicated.

Chapter 8

Recurrences

This chapter is concerned with integer sequences, and especially those that are defined recursively. Much of the material is about how to find closed-form formulas for recursively defined integer sequences, and in particular for arithmetic and geometric sequences.

8.1 Chapter 8 Definitions and Notation

8.1.1 Notation

F_n: The nth Fibonacci number.

$a_1, a_2, a_3, \ldots, a_n, \ldots$: A generic integer sequence.

8.1.2 Definitions

Fibonacci numbers: Collectively these form an integer sequence defined by the recurrence $F_n = F_{n-1} + F_{n-2}$ with initial values $1, 1$.

integer sequence: Integers listed in some order. Technically, an integer sequence is a function $s : \mathbb{N} \to \mathbb{Z}$ or $s : \mathbb{W} \to \mathbb{Z}$, but rarely do we think of a sequence in this manner.

closed-form formula: Also called a closed form, this is a rule for producing the nth term of a sequence given only the number n.

recurrence relation: Also called a recurrence, this is a rule for generating more terms of a sequence by knowing only some of the previous terms.

recursion: The process of using a recurrence.

explicit formula: A closed-form formula, in contrast to the implicit expression of a recurrence.

recurrence: A statement of the form $a_n = ($ some stuff, some of which involves $a_{\text{smaller than } n})$.

recursing: The process of iterating a recurrence.

arithmetic sequence: A sequence with constant differences.

linear recurrence relation: A recurrence relation where none of the a_{n-j} terms are raised to powers other than 1 and none are multiplied by each other.

homogeneous recurrence relation: A recurrence relation that evaluates to 0 when 0 is plugged in to the a_j on the right-hand side.

constant coefficients: A recurrence relation with coefficients of a_j terms that are not variable.

linear homogeneous recurrence relation with constant coefficients: A recurrence with the form $a_n = c_1 a_{n-1} + c_2 a_{n-2} + \cdots + c_k a_{n-k}$.

geometric sequence: A sequence where consecutive terms are related by multiplication by a constant.

8.2   Chapter 8 Facts and Theorems

Fibonacci facts:

* ✿ For $n \in \mathbb{N}$, $1 + \displaystyle\sum_{j=1}^{n} F_j = F_{n+2}$.

* ✿ For $n \in \mathbb{N}$, $F_{n-1} F_{n+1} = (F_n)^2 + (-1)^n$.

A special case.   Suppose you know some sequence a_n, with its initial terms and recurrence relation and closed form, and you encounter a sequence b_n. You intuit (and then prove) that b_n has the same recurrence relation as a_n and then discover that b_n has the same initial terms as a_n. That means that $b_n = a_n$, so it has the same closed form!

How to find a closed form for a recurrence relation of the form $a_n = a_{n-1} + p_k(n)$, $a_0 = c_0$:

1. Check to see whether the sequence has constant differences. If so, then the closed form is $a_n = c + dn$.

2. The kth differences of the corresponding sequence will be constant, so figure out k. (It is one more than the highest power in the polynomial.)

3. Use $a_0, a_1, a_2, \ldots, a_k$ to get c_0 and k linear equations in k unknowns.

4. Solve these by hand, either using high-school algebra or linear algebra. Or feed them to a computer-algebra system.

5. The k unknowns are $c_1, \ldots c_k$. The closed form is $a_n = c_0 + c_1 n + c_2 n^2 + \cdots + c_k n^k$.

How to find a closed form for a recurrence relation of the form $a_n = c_1 a_{n-1} + c_2 a_{n-2} + \ldots + c_k a_{n-k}$ (under a mild constraint to be revealed in step 5):

1. Rewrite $a_n = c_1 a_{n-1} + c_2 a_{n-2} + \cdots + c_k a_{n-k}$ as $x^n = c_1 x^{n-1} + c_2 x^{n-2} + \cdots + c_k x^{n-k}$.

2. Every term has at least an x^{n-k} in it, so divide through to get $x^{n-(n-k)} = x^k = c_1 x^{k-1} + c_2 x^{k-2} + \cdots + c_k$. This is called the *characteristic equation*.

3. Move all the terms to one side, as in $x^k - c_1 x^{k-1} - c_2 x^{k-2} - \cdots - c_k = 0$, to obtain a polynomial that we have some chance of factoring.

4. Find the roots $r_1, \ldots, r_k$ of this baby. This can be done by factoring into linear terms $(x - r_1) \cdots (x - r_k) = 0$ or by getting a computer or calculator to produce the roots.

5. Hope that $r_1, \ldots, r_k$ are all different and all real, because otherwise the rest of this algorithm won't apply. (By the way, none of the r_j are 0 because if some $r_j = 0$, then we could factor x out of the characteristic equation, and that would have been canceled in step 2.)

6. Write out the closed-form equation $a_n = q_1 r_1^n + q_2 r_2^n + \cdots + q_k r_k^n$. All that's left is to figure out what the q_j are.

7. Go find the first k terms of the sequence $a_1, a_2, \ldots, a_k$ (you must have left them lying around *some*where ...). Use these to generate the k equations

$$a_1 = q_1 r_1^1 + q_2 r_2^1 + \cdots + q_k r_k^1 = q_1 r_1 + q_2 r_2 + \cdots + q_k r_k,$$
$$a_2 = q_1 r_1^2 + q_2 r_2^2 + \cdots + q_k r_k^2,$$

$$\vdots \quad \vdots \quad \vdots$$

$$a_k = q_1 r_1^k + q_2 r_2^k + \cdots + q_k r_k^k,$$

and if you happen to have an a_0 lying around, you get a bonus simple equation

$$a_0 = q_1 r_1^0 + q_2 r_2^0 + \cdots + q_k r_k^0 = q_1 + q_2 + \ldots q_k.$$

8. If you don't happen to have an a_0 lying around, figure out what value you can assign to a_0 that is consistent with your recurrence and initial values. That gives you the bonus simple equation.

9. Use lots of high-school-algebra symbolic manipulation to determine $q_1, q_2, \ldots, q_k$ from the equations you generated in steps 7 and 8.

10. Plug these values back into $a_n = q_1 r_1^n + q_2 r_2^n + \cdots + q_k r_k^n$ and rejoice in your closed-form formula.

Lemma 8.8.6. The formula $a_n = r_j^n$ is a closed form for the recurrence $a_n = c_1 a_{n-1} + c_2 a_{n-2} + \cdots + c_k a_{n-k} \iff r_j$ is a root of a_n's characteristic equation.

8.3 Chapter 8 Proof Techniques: Showing a Closed Form for a Recurrence Is Correct

How to prove that a closed form for a recurrence is correct: Suppose that we have a sequence with initial values $a_1, a_2, a_3, a_4, \ldots$, a recursive definition $a_n =$ (some stuff, some of which involves $a_{\text{smaller than } n}$), and a formula $a_n =$ (a function of n).

1. Decide to proceed by induction.

2. Check base cases by plugging $n = 1, 2, 3$ into (the given function of n); hopefully the result will be a_1, a_2, a_3.

3. State the inductive hypothesis: for $n \le k$, $a_n = $ (the proposed function of n).

4. Consider the $(k+1)$st term, a_{k+1}.

5. Rewrite the recurrence by substituting $k + 1$ in for each copy of n, obtaining something like $a_{k+1} = $ (some stuff, some of which involves $a_{\text{smaller than } k+1}$).

6. For every $a_{\text{smaller than } k+1}$ in the rewritten recurrence, use the inductive hypothesis to substitute the formula for (the given function of smaller than $k + 1$).

7. Do some symbolic manipulations so that the rewritten recurrence becomes a function of $k + 1$.

8. Notice that you now have $a_{k+1} = $ (the given function of $k + 1$) and therefore you're done.

8.4   Some Straightforward Examples of Chapter 8 Ideas

An example of generating terms from a recurrence.   Consider the recurrence $a_1 = 1, a_2 = 1, a_n = -a_{n-1}a_{n-2} + n$.
We already have a_1, a_2, so let's start with $n = 3$.
$a_3 = -a_2 a_1 + 3 = -1 \cdot 1 + 3 = 2$. Keep going.
$a_4 = -a_3 a_2 + 4 = -2 \cdot 1 + 4 = 2$.
$a_5 = -a_4 a_3 + 5 = -2 \cdot 2 + 5 = 1$.
$a_6 = -a_5 a_4 + 6 = -1 \cdot 2 + 6 = 4$.
$a_7 = -4 \cdot 1 + 7 = 3$.
$a_8 = -3 \cdot 4 + 8 = -4$.
$a_9 = -(-4)3 + 9 = 21$.
$a_{10} = -21(-4) + 10 = 94$.
Our first 10 terms are thus $1, 1, 2, 2, 1, 2, 3, -4, 21, 94$.

An example of getting a closed form for an arithmetic sequence.   The recurrence $a_n = a_{n-1} - 9n + 5, a_0 = 1$ is arithmetic with a first-degree polynomial, so we expect that second differences will be constant. The first five terms of the sequence are $1, -3, -16, -38, -69$. We compute first and second differences.

$$
\begin{array}{ccccccc}
1 & & -3 & & -16 & & -38 & & -69 & \cdots \\
& -4 & & -13 & & -22 & & -31 & \\
& & -9 & & -9 & & -9 & &
\end{array}
$$

As expected, the second differences are constant. Therefore, the closed form will be $a_n = c + dn + fn^2$, and we need to determine c, d, and f.

When $n = 0$, we have $a_0 = 1 = 1 + d0 + f0^2$, so we know $c = 1$.

When $n = 1$, we have $a_1 = -3 = 1 + d + f$, or $d + f = -4$ or $f = -4 - d$.

When $n = 2$, we have $a_2 = -16 = 1 + 2d + 4f$, or $2d + 4f = -15$.

We now have two linear equations and two unknowns, and using high-school algebra, we can solve for d and f.

Substituting in for f, we get $2d + 4(-4 - d) = 2d - 16 - 4d = -2d - 16 = -15$, or $-2d = 1$, so $d = \frac{1}{2}$ and $f = -4 - \frac{1}{2} = -\frac{9}{2}$.

Thus, our closed form is $a_n = 1 + \frac{1}{2}n - \frac{9}{2}n^2$. Let's check and make sure this is correct:

$a_1 = 1 + \frac{1}{2} - \frac{9}{2} = 1 - 4 = -3$. Yup.

$a_2 = 1 + \frac{1}{2} \cdot 2 - \frac{9}{2} \cdot 4 = 1 + 1 - 18 = -16$. Yup. $a_3 = 1 + \frac{1}{2} \cdot 3 - \frac{9}{2} \cdot 9 = 1 + \frac{3-81}{2} = 1 - 39 = -38$. Looks like we're good.

An example of getting a closed form for a geometric sequence. The recurrence

$a_n = 2a_{n-1} + a_{n-2} - 2a_{n-3}, a_0 = 1, a_1 = 0, a_2 = -1$ is geometric. We begin by finding the characteristic polynomial: first convert as to xs and subscripts to superscripts, as in

$x^n = 2x^{n-1} + x^{n-2} - 2x^{n-3}$, then cancel extra factors of x (here, x^{n-3}), as in

$x^3 = 2x^2 + x - 2$. Rewriting so we can factor gives us

$x^3 - 2x^2 - x + 2 = 0$, and factoring gives us

$(x - 1)(x + 1)(x - 2) = 0$, so the roots are

$x = \pm 1, x = 2$.

This corresponds to the closed-form equation $a_n = q_1(1)^n + q_2(-1)^n + q_3(2)^n$, which generates the equations

$a_0 = 1 = q_1 + q_2 + q_3$

$a_1 = 0 = q_1 - q_2 + 2q_3$ (notice that this tells us $q_2 = q_1 + 2q_3$)

$a_2 = -1 = q_1 + q_2 + 4q_3$. Plugging our equation for q_2 into the first and third equations gives us

$1 = q_1 + q_1 + 2q_3 + q_3 = 2q_1 + 3q_3$ or $2q_1 = 1 - 3q_3$
and
$-1 = q_1 + q_1 + 2q_3 + 4q_3 = 2q_1 + 6q_3$ or $2q_1 = -1 - 6q_3$, from which we
can conclude that
$1 - 3q_3 = -1 - 6q_3$ or $2 = -3q_3$ or $q_3 = \frac{-2}{3}$.
From this, we have $2q_1 = 1 - 3\frac{-2}{3} = 3$ so $q_1 = \frac{3}{2}$.
Then, $q_2 = \frac{3}{2} + 2\frac{-2}{3} = \frac{1}{6}$. Our closed form is then
$a_n = \frac{3}{2}(1)^n + \frac{1}{6}(-1)^n + \frac{-2}{3}(2)^n$. As a quick check on our arithmetic, we
note that $a_0 = 1 = q_1 + q_2 + q_3$ does indeed hold.

8.5   More Problems for Chapter 8

Those solutions that model a formal write-up (such as one might hand in
for homework) are to Problems 1 and 10.

1. Dandelions reproduce very quickly, as anyone who maintains a
   lawn knows. In fact, did you know that on any given day, if you
   went to your lawn and counted the dandelions, then the next day
   twice as many *new* dandelions will have emerged from the ground?
   Luckily, dandelions die after two days, so that helps keep the numbers down. Still, if on Day 0 you had 1 dandelion, then on Day 1
   you would have 3 dandelions, on Day 2 you'd have 8 dandelions,
   and then on Day 3 you'd have 22 dandelions.

   (a) Write a recurrence equation for d_n = the number of dande-
       lions on Day n.
   (b) Find a closed-form formula for d_n.

2. Generate the first 30 terms of the sequence $a_n = a_{n-1} + a_{n-2} - a_{n-3}$,
   $a_0 = 0, a_1 = 1, a_2 = 1$.

3. Suppose that $a_n = (-4)^n$, $a_n = 1$, and $a_n = 2^n$ are all closed forms
   for the same recurrence. Find a recurrence that fits this criterion
   and verify that it really does work for all three closed forms.

4. Consider the sequence $1, 3, 4, 7, 11, 18, 29, \ldots$.

   (a) Find a recurrence that L_n satisfies.
   (b) Prove that $L_n = F_{n-1} + F_{n+1}$.

5. Find a closed-form formula for the sequence $a_0 = -1, a_n = a_{n-1} + 3n + 1$.

6. Consider the recurrence relation $a_n = 3a_{n-1} - a_{n-2}$ with $a_0 = 0, a_1 = 1$. Generate some terms, make a conjecture as to what sequence this is, try to find the closed form, and try to explain what is going on here.

7. Consider the sequence $5, -3, 5, -3, 5, -3, 5, -3, 5, -3, 5, \ldots$. Find a recurrence for this sequence, and find two more (different) sequences that satisfy that recurrence.

8. Find a closed form for the sequence defined by the recurrence $a_n = -a_{n-1}a_{n-2} + 2, a_0 = 1, a_1 = 1$. How do things change if $a_0 = 0$, $a_1 = 0$?

9. Here is a characteristic equation: $x^5 + 4x^3 - 3x^2 - 1 = 0$. What is the associated recurrence?

10. Find a closed-form formula for the sequence $a_0 = 1, a_n = a_{n-1} + n^2 - 2n$.

8.6 More Solutions for Chapter 8

1. Dandelions reproduce very quickly, as anyone who maintains a lawn knows. In fact, did you know that on any given day, if you went to your lawn and counted the dandelions, then the next day twice as many *new* dandelions will have emerged from the ground? Luckily, dandelions die after two days, so that helps keep the numbers down. Still, if on Day 0 you had 1 dandelion, then on Day 1 you would have 3 dandelions, on Day 2 you'd have 8 dandelions, and then on Day 3 you'd have 22 dandelions.

   (a) Write a recurrence equation for d_n = the number of dandelions on Day n.

   (b) Find a closed-form formula for d_n.

(a) Let's see what happens day by day.
Day 0: 1
Day 1: $1 + 2 = 3$ (one old, two new)
Day 2: $3 + 6 - 1 = 8$ (three old, six new, and the two-day-old one dies)
Day 3: $8 + 16 - 2 = 22$ (eight old, sixteen new, and the two two-day-old dandelions die)
Day 4: $22 + 44 - 6 = 60$. It looks like we have
$d_n = 3d_{n-1} - 2d_{n-3}$.

(b) Characteristic polynomial: $x^n = 3x^{n-1} - 2x^{n-3}$ becomes
$x^3 = 3x^2 - 2$ becomes
$x^3 - 3x^2 + 2 = 0$ which has roots
$x = 1, x = 1 \pm \sqrt{3}$.
Now we have $d_n = q_1(1)^n + q_2(1 + \sqrt{3})^n + q_3(1 - \sqrt{3})^n$.
For low values of n, we have
$d_0 = 1 = q_1 + q_2 + q_3$
$d_1 = 3 = q_1 + q_2(1 + \sqrt{3}) + q_3(1 - \sqrt{3})$
$d_2 = 8 = q_1 + q_2(1 + \sqrt{3})^2 + q_3(1 - \sqrt{3})^2 = q_1 + q_2(4 + 2\sqrt{3}) + q_3(4 - 2\sqrt{3})$.
Multiply the d_1 equation by -2 and add it to the d_2 equation to get
$2 = -q_1 + 2q_2 + 2q_3$.
Multiply the d_0 equation by -2 and add it to $2 = -q_1 + 2q_2 + 2q_3$ to get $0 = -3q_1$ or $q_1 = 0$.
Rewrite the first two original equations:
$1 = q_2 + q_3$

$3 = q_2(1 + \sqrt{3}) + q_3(1 - \sqrt{3})$.

Now use $q_2 = 1 - q_3$ to get $3 = (1 - q_3)(1 + \sqrt{3}) + q_3(1 - \sqrt{3}) = 1 + \sqrt{3} - 2\sqrt{3}q_3$ so that $q_3 = \frac{2-\sqrt{3}}{2\sqrt{3}}$

and $q_2 = 1 - \frac{2-\sqrt{3}}{2\sqrt{3}} = \frac{1}{2} + \frac{1}{\sqrt{3}}$.

Our final formula is $d_n = \left(\frac{1}{2} + \frac{1}{\sqrt{3}} \right) (1 + \sqrt{3})^n + \frac{2-\sqrt{3}}{2\sqrt{3}} (1 - \sqrt{3})^n$.

Yuck!

2. Generate the first 30 terms of the sequence $a_n = a_{n-1} + a_{n-2} - a_{n-3}$, $a_0 = 0, a_1 = 1, a_2 = 1$.

$1, 1, 2, 2, 3, 3, 4, 4, 5, 5, 6, 6, 7, 7, 8, 8, 9, 9, 10, 10, 11, 11,$
$12, 12, 13, 13, 14, 14, 15, 15, 16, 16, 17, 17, 18, 18, 19, 19, 20, 20.$ Ha!

3. Suppose that $a_n = (-4)^n$, $a_n = 1$, and $a_n = 2^n$ are all closed forms for the same recurrence. Find a recurrence that fits this criterion and verify that it really does work for all three closed forms.

This criterion suggests we have a geometric sequence whose characteristic polynomial has roots $r_1 = -4, r_2 = 1, r_3 = 2$, and that from various initial conditions we get some of the $q_i = 0$. The characteristic polynomial would be $(x + 4)(x - 1)(x - 2) = x^3 + x^2 - 10x + 8 = 0$, or $x^3 = -x^2 + 10x - 8$, which corresponds to a recurrence $a_n = -a_{n-1} + 10a_{n-2} - 8a_{n-3}$.

Can we find initial conditions that correspond to these three sequences? We *should* be able to:

If $a_n = (-4)^n$, then $a_0 = 1, a_1 = -4, a_2 = 16, a_3 = -64$. Is it true that $-64 = -16 + 10(-4) - 8$? Yes!

If $a_n = 1$, then $a_0 = 1, a_1 = 1, a_2 = 1, a_3 = 1$. Is it true that $1 = -1 + 10(1) - 8$? Yes!

If $a_n = 2^n$, then $a_0 = 1, a_1 = 2, a_2 = 4, a_3 = 8$. Is it true that $8 = -4 + 10(2) - 8$? Yes!

4. Consider the sequence $1, 3, 4, 7, 11, 18, 29, \ldots$.

    (a) Find a recurrence that L_n satisfies.

    (b) Prove that $L_n = F_{n-1} + F_{n+1}$.

(a) $L_n = L_{n-1} + L_{n-2}$.

(b) We'll do this by induction. First we need to figure out how to make the indices of the two sequences match (that is, do we want

$L_0 = 1$ or $L_1 = 1$?). Note that $3 = 1 + 2$, so the second term of the L sequence is the sum of the first and third Fibonacci numbers. So we have $L_2 = F_1 + F_3$. We'll check another couple to be sure: $L_3 = 4 = 1 + 3 = F_2 + F_4$ and $L_4 = 7 = 2 + 5 = F_3 + F_5$. These will function as our base cases.

Now assume that our statement holds for values of $n \leq k$. $L_{k+1} = L_k + L_{k-1}$ and the inductive hypothesis gives us $L_k = F_{k-1} + F_{k+1}$, $L_{k-1} = F_{k-2} + F_k$, so $L_{k+1} = F_{k-1} + F_{k+1} + F_{k-2} + F_k = (F_{k-1} + F_{k-2}) + (F_{k+1} + F_k) = F_k + F_{k+2}$ as desired.

5. Find a closed-form formula for the sequence $a_0 = -1, a_n = a_{n-1} + 3n + 1$.

   This is an arithmetic sequence, so we use the technique of finite differences.
   Sequence: $-1, 3, 10, 20, 33, 49, 68, 90, 115, 143, 174\ldots$
   First differences: $4, 7, 10, 13, 16, 19, 22, 25, 28, 31\ldots$
   Second differences: $3, 3, 3, 3, 3, 3, 3, 3, 3\ldots$
   Closed form is generically $a_n = c + dn + fn^2$.
   When $n = 0$, we have $\quad 1 = c$.
   When $n = 1$, we have $3 = c + d + f$ or $4 = d + f$ or $f = 4 - d$.
   When $n = 2$, we have $10 = c + 2d + 4f$ or $11 = 2d + 4(4 - d) = 16 - 2d$.
   Then $d = \frac{5}{2}$ and $f = \frac{3}{2}$, so
   $a_n = -1 + \frac{5}{2}n + \frac{3}{2}n^2$.

6. Consider the recurrence relation $a_n = 3a_{n-1} - a_{n-2}$ with $a_0 = 0, a_1 = 1$. Generate some terms, make a conjecture as to what sequence this is, try to find the closed form, and try to explain what is going on here.

   Terms: $0, 1, 3, 8, 21, 55, \ldots$
   I gotta say, this looks like every second Fibonacci number. I wonder what kind of formula that will give us?
   Let's start with $F_n = F_{n-1} + F_{n-2}$. We want only the even ones, so F_{2k}. Those have the form $F_{2k} = F_{2k-1} + F_{2k-2}$. But $2k - 1$ is odd, so let's break that down further. (Uh-oh. This might never end.)
   $F_{2k} = (F_{2k-2} + F_{2k-3}) + F_{2k-2}$. But wait, $F_{2k-2} = F_{2k-3} + F_{2k-4}$ can be written as $F_{2k-3} = F_{2k-2} - F_{2k-4}$, which means we get
   $F_{2k} = (F_{2k-2} + F_{2k-2} - F_{2k-4}) + F_{2k-2} = 3F_{2k-2} - F_{2k-4}$. And *that* can be rewritten as

$F_{2(k)} = 3F_{2(k-1)} - F_{2(k-2)}$, or, ditching the 2s, as $a_n = 3a_{n-1} - a_{n-2}$. So that's what's going on—this recurrence is the same as for every second Fibonacci number.

Oh, we have to find a closed form. Okay.
Characteristic equation: $x^n = 3x^{n-1} - x^{n-2}$ becomes $x^2 = 3x - 1$ or $x^2 - 3x + 1 = 0$.
The roots of this equation are $\frac{3-\sqrt{5}}{2}, \frac{3+\sqrt{5}}{2}$.
Therefore we have $a_n = q_1 \left(\frac{3-\sqrt{5}}{2}\right)^n + q_2 \left(\frac{3+\sqrt{5}}{2}\right)^n$.
For low values of n, we have
$a_0 = 0 = q_1 + q_2$
$a_1 = 1 = q_1 \left(\frac{3-\sqrt{5}}{2}\right) + q_2 \left(\frac{3+\sqrt{5}}{2}\right)$.
Now $q_2 = -q_1$, so $1 = q_1 \left(\frac{3-\sqrt{5}}{2}\right) - q_1 \left(\frac{3+\sqrt{5}}{2}\right) = q_1 \left(-4\sqrt{5}\right)$ so
$q_1 = \frac{-1}{4\sqrt{5}}$ and $q_2 = \frac{1}{4\sqrt{5}}$ for a final closed-form formula of
$a_n = \left(\frac{-1}{4\sqrt{5}}\right) \left(\frac{3-\sqrt{5}}{2}\right)^n + \left(\frac{1}{4\sqrt{5}}\right) \left(\frac{3+\sqrt{5}}{2}\right)^n$.

7. Consider the sequence $5, -3, 5, -3, 5, -3, 5, -3, 5, -3, 5\ldots$. Find a recurrence for this sequence, and find two more (different) sequences that satisfy that recurrence.

One recurrence for this sequence is $a_n = -a_{n-1} + 2$.
Starting with $a_0 = 1$, we get $1, 1, 1, 1, 1, 1, 1, 1, 1, 1, 1\ldots$.
Starting with $a_0 = 0$, we get $0, 2, 0, 2, 0, 2, 0, 2, 0, 2, 0\ldots$.
(And starting with $a_0 = 2$, we get $2, 0, 2, 0, 2, 0, 2, 0, 2, 0, 2\ldots$!)

8. Find a closed form for the sequence defined by the recurrence $a_n = -a_{n-1}a_{n-2} + 2, a_0 = 1, a_1 = 1$. How do things change if $a_0 = 0$, $a_1 = 0$?

The sequence is $1, 1, 1, 1, 1, 1, 1, 1, 1, 1, 1\ldots$, so the closed form is $a_n = 1$.

If we change the initial conditions, we get the sequence $0, 0, 2, 2, -2, 6, 14, -82, 1150, 94302, -108447298\ldots$, which is (at this writing) not even in the OEIS.

9. Here is a characteristic equation: $x^5 + 4x^3 - 3x^2 - 1 = 0$. What is the associated recurrence?

First we isolate the highest term to get
$x^5 = -4x^3 + 3x^2 + 1$.
Next we multiply through by x^{n-5} to get

$$x^n = -4x^{n-2} + 3x^{n-3} + x^{n-5}.$$

Finally, we replace x-superscript with a-subscript to get

$$a_n = -4a_{n-2} + 3a_{n-3} + a_{n-5}.$$

10. Find a closed-form formula for the sequence $a_0 = 1, a_n = a_{n-1} + n^2 - 2n$.

This is an arithmetic sequence, so we use the technique of finite differences.

Sequence: $1, 0, 0, 3, 11, 26, 50, 85, 133, 196, 276 \ldots$

First differences: $-1, 0, 3, 8, 15, 24, 35, 48, 63, 80 \ldots$

Second differences: $1, 3, 5, 7, 9, 11, 13, 15, 17 \ldots$

Third differences: $2, 2, 2, 2, 2, 2, 2, 2 \ldots$

Closed form is generically $a_n = c + dn + fn^2 + gn^3$.

When $n = 0$, we have $1 = c$.

When $n = 1$, we have $0 = c + d + f + g$ or $g = -1 - d - f$.

When $n = 2$, we have $0 = c + 2d + 4f + 8g$ or $0 = 1 + 2d + 4f + 8(-1 - d - f) = -7 - 6d - 4f$ or $f = \frac{-7-6d}{4}$.

When $n = 3$, we have $3 = c + 3d + 9f + 27g$ or $3 = 1 + 3d + 9(\frac{-7-6d}{4}) + 27(-1 - d - \frac{-7-6d}{4}) = 3d + \frac{11}{2}$.

Then $d = \frac{-5}{6}$ and $f = \frac{-1}{2}$ and $g = \frac{1}{3}$, so

$$a_n = 1 + \frac{-5}{6}n + \frac{-1}{2}n^2 + \frac{1}{3}n^3.$$

Chapter 9 🦆🦆🦆🦆🦆🦆🦆🦆🦆

Cutting Up Food:
Counting and Geometry

This chapter concentrates on one problem: Into how many regions do a bunch of $(d-1)$-dimensional knives slice d-dimensional space? That's kind of brain-break-y, so we start in two dimensions and work our way up.

9.1 Chapter 9 Definitions and Notation

9.1.1 Notation

p_n: The maximum number of pieces of pizza obtainable using n cuts.

y_n: The maximum number of pieces of yam obtainable using n cuts.

s_n: The maximum number of pieces of spaghetti obtainable using n cuts.

$h_{n,k}$: The maximum number of pieces of k-dimensional hyperbeet obtainable using n hypercuts.

9.1.2 Definitions

general position: In two dimensions, a placement of lines so that no two are parallel and no three intersect at a single point. In three dimensions, no two cuts may be parallel, no three cuts may meet in a line, and no four cuts may meet at a point. In k dimensions, no two cuts should be parallel, no three cuts should intersect in a $(k-1)$-dimensional cut, ..., no k cuts should intersect in a line, and no $k+1$ cuts should intersect at a point.

hyperbeet: A k-dimensional root vegetable.

hypercut: A cut in k-dimensional space, so a $(k-1)$-dimensional flat space.

9.2   Chapter 9 Facts and Theorems

Pizza summary.   In order to maximize the number of pieces of pizza, we want the cuts to be placed in general position.

Recursive approach to p_n: Assume that we have $n-1$ lines placed such that the maximum number of regions p_{n-1} is attained. From geometry we know that two lines that are not parallel must intersect. Thus, as long as we place an nth line so that it is not parallel to any of the already placed $n-1$ lines, it will intersect each of them. This creates n new regions (one "above" each line intersected and one "below" the lowest line), and therefore $p_n = p_{n-1} + n$. (We know that this recurrence, together with the initial values of $p_0 = 1, p_1 = 2$, produces the closed form $p_n = \frac{n^2}{2} + \frac{n}{2} + 1$.)

Direct counting approach to p_n: Imagine that there are n lines dividing the plane into a maximum number of regions (and thus in general position). Then, imagine a glowing line outside of the area that contains all the intersections (outside the pizza, as it were). The glowing line intersects $n+1$ regions because it is cut in n places. Now, sweep the glowing line across all the intersections; each time the glowing line sweeps across an intersection, it enters a new region. If we count the number of intersections, we will then know how many regions there are. Every pair of lines crosses, so there are exactly $\binom{n}{2}$ intersections. Thus, the total number of regions is $p_n = \binom{n}{2} + n + 1$.

Yam summary.   We are arranging planes in three-dimensional space and counting the number of regions that result. To achieve a maximum number of regions, the planes must be in general position.

Recursive approach to y_n: Suppose we have $n-1$ planes placed in general position in three-dimensional space, so that y_{n-1} regions are present. Let us count the number of new regions created when an nth plane is added (in general position). This is equal to the number of regions through which the nth plane passes. Intersecting this plane is a network of lines, separating the plane into regions. Those lines are cross sections of the $n-1$ other planes that carve up the 3-dimensional space, and the regions are cross sections of the solid regions through which the nth plane passes. Thus, the number of regions is the corresponding pizza number p_{n-1}, and therefore $y_n = y_{n-1} + p_{n-1}$.

Direct counting approach to y_n: Move a softly glowing plane through all the intersections made by n planes in general position. When the glowing plane is outside all the intersections (outside the yam, so to speak), the glowing plane cuts through $\binom{n}{2} + n + 1$ three-dimensional regions. As the glowing plane passes an intersection, it moves into a new region, and so there are as many additional regions as intersections. Because the planes are in general position, every three of them intersect at a point—so, there are $\binom{n}{3}$ such intersections. In total, we have $\binom{n}{3} + \binom{n}{2} + n + 1$ regions.

Hyperbeet summary. We will denote the maximal number of hyperchunks obtainable when cutting a k-dimensional hyperbeet with n hypercuts by $h_{n,k}$.

Recursive approach to $h_{n,k}$: Suppose we have $n - 1$ hypercuts placed in general position in k-dimensional space. We must count the number of new regions created when an nth hypercut is added in general position. This is equal to the number of regions through which the nth hypercut passes. Intersecting this hypercut is a network of $(k - 2)$-dimensional cuts in general position that separate the hypercut into $(k - 1)$-dimensional hyperregions. Thus we have $h_{n-1,k-1}$ hyperregions represented. Each of these hyperregions corresponds to a new region created by the nth hypercut cleaving an old region in two. So, we have the $h_{n-1,k}$ regions created by the first $n - 1$ hypercuts, and to this we add the $h_{n-1,k-1}$ regions created by the nth hypercut. Thus, $h_{n,k} = h_{n-1,k} + h_{n-1,k-1}$.

Direct counting approach to $h_{n,k}$: We move a softly glowing hypercut through all the intersections made by n hypercuts in general position. When the glowing hypercut is outside all the intersections (i.e., outside the hyperbeet), it cuts through $\binom{n}{k-1} + \cdots + \binom{n}{3} + \binom{n}{2} + \binom{n}{1} + \binom{n}{0}$ k-dimensional regions because that is how many $(k - 1)$-dimensional regions are demarcated by the n cross sections of hypercuts on the glowing hypercut itself, and each such $(k - 1)$-dimensional region bounds an unbounded k-dimensional region. Now we move the glowing hypercut through the hyperbeet. As the glowing hypercut passes an intersection, it moves into a new region, and so there are as many additional regions as intersections. Because the hypercuts are in general position, every k of them intersect at a point—so, there are $\binom{n}{k}$ intersections. In total, we have $\binom{n}{k} + \cdots + \binom{n}{3} + \binom{n}{2} + \binom{n}{1} + \binom{n}{0}$ regions.

9.3 More Problems for Chapter 9

Those solutions that model a formal write-up (such as one might hand in for homework) are to Problems 3 and 5.

1. Let f_n be the maximum number of regions of 4-dimensional space that are cut up by n 3-dimensional cuts. What are f_0, f_1, f_2, f_3, f_4? And why?

2. If you cut a configuration with f_4 pieces with an additional cut, how many new pieces can you get?

3. Determine and explain a recurrence relation for f_n.

4. Determine and explain a closed form for f_n.

5. Use induction to prove that your closed form from Problem 4 is the correct closed form for your recurrence from Problem 3.

9.4   More Solutions for Chapter 9

1. Let f_n be the maximum number of regions of 4-dimensional space that are cut up by n 3-dimensional cuts. What are f_0, f_1, f_2, f_3, f_4? And why?

   $f_0 = 1, f_1 = 2, f_2 = 4, f_3 = 8, f_4 = 16$ because there are four perpendicular directions and each cut can be perpendicular to the previous cuts—for these values of n.

2. If you cut a configuration with f_4 pieces with an additional cut, how many new pieces can you get?

   There can be as many as 15 new pieces, which means that the additional cut passes through all but one of the existing regions. We know this because ... see the answer for the next question!

3. Determine and explain a recurrence relation for f_n.

   $f_n = f_{n-1} + y_{n-1}$. That's because when we look at one of the n cuts, it's really a yam with a maximum number of planes passing through it, and there are $n - 1$ such planes formed by intersecting the 3-dimensional cuts with the nth cut (yam). Each of those y_{n-1} regions of the yam represents a cutting-in-two of the region it passes through in 4-dimensional space. So we add y_{n-1} to the f_{n-1} regions we already had.

4. Determine and explain a closed form for f_n.

   $$f_n = \binom{n}{4} + \binom{n}{3} + \binom{n}{2} + \binom{n}{1} + \binom{n}{0}.$$

   If we have a maximum number of regions, then the cuts are in general position. That means that any four cuts intersect in exactly one point. Consider an extra cut; if we sweep it across the four-dimensional space, the number of point-intersections it crosses will be $\binom{n}{4}$. Every time it passes an intersection, it crosses into a new region. And after it has passed through all of the intersections, it has $\binom{n}{2} + \binom{n}{1} + \binom{n}{0}$ regions passing through it because that's the number of regions *on* it of lower dimension.

5. Use induction to prove that your closed form from Problem 4 is the correct closed form for your recurrence from Problem 3.

   Here are some base cases:
   $n = 0$: $\binom{0}{4} + \binom{0}{3} + \binom{0}{2} + \binom{0}{1} + \binom{0}{0} = 0 + 0 + 0 + 0 + 1 = 1$.

$n = 1$: $\binom{1}{4} + \binom{1}{3} + \binom{1}{2} + \binom{1}{1} + \binom{1}{0} = 0 + 0 + 0 + 1 + 1 = 2.$

$n = 2$: $\binom{2}{4} + \binom{2}{3} + \binom{2}{2} + \binom{2}{1} + \binom{2}{0} = 0 + 0 + 1 + 2 + 1 = 4.$

$n = 3$: $\binom{3}{4} + \binom{3}{3} + \binom{3}{2} + \binom{3}{1} + \binom{3}{0} = 0 + 1 + 3 + 3 + 1 = 8.$

$n = 4$: $\binom{4}{4} + \binom{4}{3} + \binom{4}{2} + \binom{4}{1} + \binom{4}{0} = 1 + 4 + 6 + 4 + 1 = 16.$

Suppose that for $n \leq k$, we have that $f_n = \binom{n}{4} + \binom{n}{3} + \binom{n}{2} + \binom{n}{1} + \binom{n}{0}$.

Consider f_{k+1}. From our recurrence, we know that $f_{k+1} = f_k + y_k$. Using the inductive hypothesis, we have

$f_{k+1} = (\binom{k}{4} + \binom{k}{3} + \binom{k}{2} + \binom{k}{1} + \binom{k}{0}) + y_k$, and from the closed form we calculated for y_k, we have that

$f_{k+1} = (\binom{k}{4} + \binom{k}{3} + \binom{k}{2} + \binom{k}{1} + \binom{k}{0}) + (\binom{k}{3} + \binom{k}{2} + \binom{k}{1} + \binom{k}{0})$.

Rearranging terms, we have

$f_{k+1} = (\binom{k}{4} + \binom{k}{3}) + (\binom{k}{3} + \binom{k}{2}) + (\binom{k}{2} + \binom{k}{1}) + (\binom{k}{1} + \binom{k}{0}) + \binom{k}{0}$

and using our basic choice identity four times while noting that $\binom{k}{0} = 1 = \binom{k+1}{0}$, we get

$f_{k+1} = \binom{k+1}{4} + \binom{k+1}{3} + \binom{k+1}{2} + \binom{k+1}{1} + \binom{k+1}{0}$ as desired.

Part III

Theme: Graph Theory

Chapter 10 🦆🦆🦆🦆🦆🦆🦆🦆🦆

Trees

In terms of content, this chapter follows directly on Chapter 3, though the reader will certainly find more success if s/he has also learned material from some intervening chapters. The study here is of trees, which are connected graphs with no cycles, and we focus on algorithms about or that use trees.

10.1   Chapter 10 Definitions and Notation

10.1.1   Definitions

tree: A connected graph without cycles.

spanning tree: A tree that contains all the vertices of a given graph; it is the largest possible tree that is also a subgraph.

weights: Labels on the edges and/or vertices of a graph that often denote costs or distances or energies.

weighted graph: A graph labeled with weights.

greedy algorithm: An algorithm that selects the best option available at every stage.

parsimonious algorithm: An algorithm that has the goal of minimizing something, and so selects the smallest or least option available at every stage.

root: A distinguished vertex.

binary tree: A tree with a root, at most two edges growing "downward" from each vertex, and a designation of "left" or "right" for each edge.

rank: The collection of vertices that are the same distance k from the root.

complete binary tree: A binary tree with exactly two edges growing "downward" from each vertex, except for those leaves in the bottom rank.

binary decision tree: A binary tree with a two-option question associated with each node, answers to this question associated to the edges incident to that node, and data associated with the bottom-rank leaves.

binary search tree: A binary tree with a datum associated with each node such that in the ordering of the data set, the datum occurs earlier than any of the data downward and to the right and occurs later than any of the data downward and to the left.

matching: A subgraph of a graph in which every vertex has degree 1. In other words, it's a pile of edges (with vertices included).

perfect matching: A matching that includes all vertices of a graph.

backtracking: An algorithm for solving problems whose solutions can be expressed as finite sequences.

backtracking tree: A rooted tree of partial potential solutions to a problem, with rank k hosting length-k partial potential solutions.

depth-first search: A method of searching a tree that starts by going down and left until either no down/left edges remain or the searched-for item is found. When no down/left edges remain, the search goes back up until it can go down/right.

10.2   Chapter 10 Facts and Theorems

Fact from Example 4.2.5.   If a tree has n vertices, then it has $n-1$ edges.

Theorem 10.2.1.   If a connected graph G has n vertices and $n-1$ edges, then G is a tree.

Theorem 10.2.2.   If a graph G with no cycles has n vertices and $n-1$ edges, then G is a tree.

Corollary 10.2.3. If a graph G has n vertices and *fewer* than $n-1$ edges, then G is not connected.

Facty fact. Trees are the connected graphs with the least number of edges.

Theorem 10.2.5. Every tree T with at least two vertices has at least two leaves.

Algorithm for finding a spanning tree—start big:

1. Begin with a connected graph G.

2. Consider a duplicate of G and name it H.

3. Pick an edge, any edge, of H and call it e.

4. If $H \setminus e$ is connected, remove e and rename $H \setminus e$ as H; otherwise, mark e as necessary (and don't consider it again).

5. Pick any edge of H not marked as necessary and call it e. If there are no unmarked edges left, rename H as T and be done.

6. Go to step 4.

Algorithm for finding a spanning tree—start small:

1. Begin with a graph G.

2. Grab a copy of the null graph (no vertices or edges) and name it H.

3. Pick an edge, any edge, of G and call it e.

4. If $H \cup e$ is a tree, add e and its vertices to H and rename $H \cup e$ as H; otherwise, mark e as superfluous.

5. Pick an edge of G incident to H and not marked as superfluous and call it e. If there are no unmarked edges left, rename H as T and be done.

6. Go to step 4.

Kruskal's algorithm for finding a minimum-weight spanning tree:

1. Begin with a connected edge-weighted graph G. Order the edges in increasing order of weight as $e_1, \ldots, e_n$.

2. Let e_1 along with its vertices be called H and set $j = 2$.

3. If $H \cup e_j$ has no cycles, then rename $H \cup e_j$ as H; otherwise, do nothing.

4. If $|E(H)| = |V(G)| - 1$, output H as the desired tree; otherwise, do nothing.

5. Increment j by 1 and go to step 3.

Prim's algorithm for finding a minimum-weight spanning tree:

1. Let G be a connected edge-weighted graph. Find the edges of least weight in G and pick one of them. Name it *ankka*.

2. Let *ankka* along with its vertices be called P.

3. Look at all the edges of G that have exactly one vertex in P. Among these, pick one of least weight and name it e. If $P \cup e$ has no cycles, then add e and its other vertex to P, and rename $P \cup e$ as P. Otherwise, mark e as *bad* so you don't look at it again.

4. If $V(P) = V(G)$, then we're done and output P. Otherwise, go to step 3.

Greedy (or parsimonious) algorithms in outline. A greedy/parsimonious algorithm tries to find the best possible option at each iteration. Usually it proceeds through a graph in a systematic way.

1. Select a place to start.

2. Look nearby.

3. See which nearby things may help to solve the problem.

4. Check to see whether the problem has been solved.

5. If the problem is solved, be done; if not, go to step 2.

Backtracking algorithms in outline. A backtracking algorithm constructs a solution that is a finite sequence.

1. Start with an empty list.

2. Append the first possible solution element.

3. Check to see whether this partial solution is valid.

4. If so, check to see whether the solution is complete; if so, terminate; if not, continue.

5. If so, append the earliest possible sequence element and go to step 3; if not, remove the last sequence element and replace it with the next possible sequence element.

10.3 Some Straightforward Examples of Chapter 10 Ideas

An example of a spanning and weighted spanning tree. In the center of Figure 10.1 is a weighted graph. At left, we have the graph with a spanning tree highlighted, but this is not a minimum-weight spanning tree; at right, the unique minimum-weight spanning tree is highlighted.

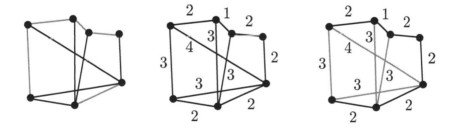

Figure 10.1. A spanning tree highlighted, a weighted graph, and a minimum-weight spanning tree highlighted.

An example of a binary search tree vs. a binary decision tree. A binary search tree stores data on all of the vertices, whereas a binary decision tree stores queries on the non-leaf vertices. We will show a binary search tree and a binary decision tree for the same data set, a collection of seven cat toys {*teal mouse, blue milk ring, catnip candy cane, paper bag, white milk ring, pink-and-purple mouse, catnip snake*}, abbreviated {tm, bmr, ccc, pb, wmr, papm, cs} for convenience.

First, Figure 10.2 shows a binary search tree that has ordered the data alphabetically.

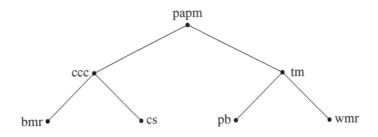

Figure 10.2. An alphabetical search tree for seven cat toys.

Next, Figure 10.3 shows a binary decision tree for identifying one of the cat toys.

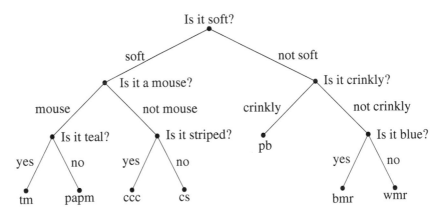

Figure 10.3. This binary decision tree helps us identify cat toys.

Example 10.5.3 rewritten.   Claim: You have an ancestor A, whose parents P_1, P_2 have a common earlier ancestor B who lived less than 1,000 years ago. We will prove this claim by contradiction and assume no such B exists.

Note that if no common earlier ancestor exists, we can model your ancestry as a complete binary tree. The root is you. In the next rank are your genetic parents and in the following rank are your four genetic grandparents, etc. But if there are two vertices in this tree that represent the same person B, then that person has two different paths of descendants that meet first in some ancestor A of yours (so there is a cycle), and A's parents are both descendants of B.

We will overestimate the average time between generations as 28 years (the mean age of a mother at first birth is about 25 now, but was much earlier a hundred or so years back, and this offsets the effect of multiple children). Therefore, there are at least 35 ranks in your ancestry tree over the last 1,000 years, for a total of at least $2 + \cdots + 2^{35} = 68{,}719{,}476{,}734$ ancestors.

On the other hand, the upper-bound estimate for the cumulative population of the world for the last 1,000 years (again using generations every 28 years) is $45{,}314{,}000{,}000$. That's a smaller number, so we have a contradiction—your ancestry tree cannot be a complete binary tree. Thus, B exists and one of your recent-ish ancestors has parents who have a common earlier ancestor.

10.4 More Problems for Chapter 10

Those solutions that model a formal write-up (such as one might hand in for homework) are to Problems 4, 5, and 8.

1. Find two different spanning trees of the graph shown at left in Figure 10.4.

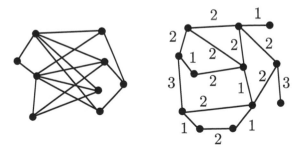

Figure 10.4. A graph and an edge-weighted graph.

2. Find two different minimum-weight spanning trees of the graph shown at right in Figure 10.4. Are there more?

3. Find, if possible, a perfect matching in each of the graphs shown in Figure 10.4.

4. Prove that for $n \geq 3$, every n-vertex tree has at most $n - 1$ leaves.

5. Create a binary search tree for the mini-dictionary {*block, black, brack, bract, brace, trace, race, ace, mace, maze, maize, baize*}.

6. Find a minimum-weight spanning tree of the graph shown at left in Figure 10.5 using Kruskal's algorithm.

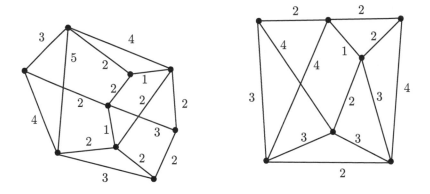

Figure 10.5. Two edge-weighted graphs.

7. Create an efficient binary decision tree for identifying members of the set {*coat, mittens, hat, scarf, duck, boots*}.

8. Prove that in any tree with at least two vertices, any two vertices are connected by a unique minimum-length path.

9. Use backtracking to find all the ways to add numbers from {1, 2, 3, 4, 5} to get 8.

10. Find a minimum-weight spanning tree of the graph shown at right in Figure 10.5 using Prim's algorithm.

10.5 More Solutions for Chapter 10

1. Find two different spanning trees of the graph shown at left in Figure 10.4.

   Figure 10.6 shows two different spanning trees of the graph shown at left in Figure 10.4.

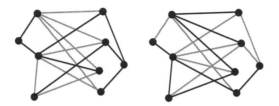

Figure 10.6. A graph underlying two of its spanning trees.

2. Find two different minimum-weight spanning trees of the graph shown at right in Figure 10.4. Are there more?

   Figure 10.7 shows all three different minimum-weight spanning trees of the graph shown at right in Figure 10.4.

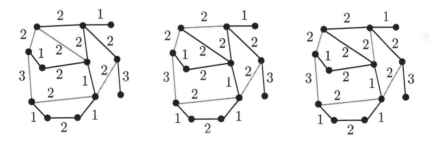

Figure 10.7. An edge-weighted graph with all of its minimum-weight spanning trees.

   You can show that these are the only three by eliminating the weight-3 edge e and then successively eliminating the highest-weight edges incident to e's vertices, etc.

3. Find, if possible, a perfect matching in each of the graphs shown in Figure 10.4.

   The left-hand graph has an odd number of vertices, so none is possible. Several are possible for the right-hand graph; the lowest-weight perfect matching has total weight 9.

4. Prove that for $n \geq 3$, every n-vertex tree has at most $n - 1$ leaves.

   We proceed by contradiction. Suppose an n-vertex tree has more than $n - 1$ leaves. In this case, it must have at least n leaves, and that means that every vertex is a leaf. However, consider the handshake lemma: The total degree must equal twice the number of edges. Here the total degree is n, and twice the number of edges is $2(n-1)$ because any n-vertex tree has $n - 1$ edges. The statement $n = 2(n-1)$ becomes $n = 2n - 2$ or $n = 2$, which violates the constraints of the theorem.

   As a side note, the star graph is an n-vertex tree with exactly $n - 1$ leaves.

5. Create a binary search tree for the mini-dictionary {*block, black, brack, bract, brace, trace, race, ace, mace, maze, maize, baize*}.

   Figure 10.8 shows such a binary search tree.

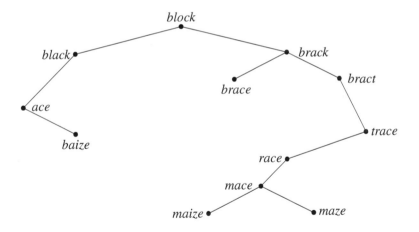

Figure 10.8. A binary search tree for a mini-dictionary.

   It was created by placing *block* at the root and then adding the remaining words to the tree in the order they were listed.

6. Find a minimum-weight spanning tree of the graph shown at left in Figure 10.5 using Kruskal's algorithm.

   Kruskal's algorithm performed on the graph shown at left in Figure 10.5 is shown at left in Figure 10.9. Edges added later are shown in lighter grey tones. There is some choice as to which edges of a given weight are added, so your solution may differ and still be correct.

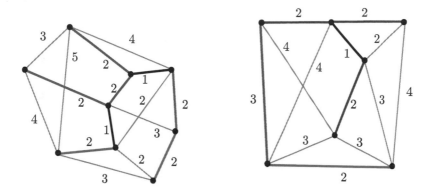

Figure 10.9. Two edge-weighted graphs with minimum spanning trees highlighted.

7. Create an efficient binary decision tree for identifying members of the set {*coat, mittens, hat, scarf, duck, boots*}.

First, note that a really terrible question is, "Is it warm?". We want to avoid questions for which all items answer *yes* (or *no*), and questions for which just one item answers *yes* (or *no*) because then we'll need more questions. Figure 10.10 gives one possible efficient tree.

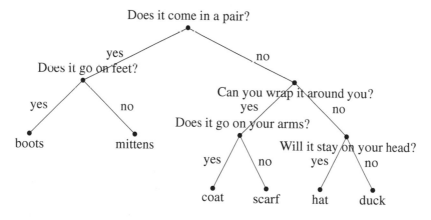

Figure 10.10. This binary decision tree helps us identify warm winter wear.

8. Prove that in any tree with at least two vertices, any two vertices are connected by a unique minimum-length path.

We proceed by contradiction. Suppose there exist two vertices v, w in a tree that are connected by at least two distinct minimum-length paths P and Q. Starting at v and heading toward w, there is some first vertex after which P, Q differ. Call this z (though it could be v itself, so then $z = v$). Continuing along P, Q, there is some first vertex after z that P, Q have in common. Call this y. Now there are two completely distinct paths from z to y, and together these form a cycle. This contradicts tree-ness.

9. Use backtracking to find all the ways to add numbers from $\{1, 2, 3, 4, 5\}$ to get 8.

1 isn't enough.
$1 + 2$ isn't enough.
$1 + 2 + 3$ isn't enough.
$1 + 2 + 3 + 4$ is too much. Go back.
$1 + 2 + 4$ isn't enough.
$1 + 2 + 4 + 5$ is too much. Go back.
$1 + 2 + 5$ is exactly right! Keep that and go back.
$1 + 3$ isn't enough.
$1 + 3 + 4$ is exactly right! Keep that and go back.
$1 + 4$ isn't enough.
$1 + 4 + 5$ is too much. Go back.
$1 + 5$ isn't enough. Go back.
2 isn't enough.
$2 + 3$ isn't enough.
$2 + 3 + 4$ is too much. Go back.
$2 + 4$ isn't enough.
$2 + 4 + 5$ is too much. Go back.
$2 + 5$ isn't enough. Go back.
3 isn't enough.
$3 + 4$ isn't enough.
$3 + 4 + 5$ is too much. Go back.
$3 + 5$ is exactly right! Keep that and go back.
4 isn't enough.
$4 + 5$ is too much. Go back.
5 isn't enough.
Report: $8 = 1 + 2 + 5 = 1 + 3 + 4 = 3 + 5$.

10. Find a minimum-weight spanning tree of the graph shown at right in Figure 10.5 using Prim's algorithm.

Prim's algorithm performed on the graph shown at right in Figure 10.5 is shown at right in Figure 10.9. Edges added later are shown in lighter grey tones. There is some choice as to which edges of a given weight are added, so your solution may differ and still be correct.

Chapter 11 🦆🦆🦆🦆🦆🦆🦆🦆🦆🦆

Euler's Formula and Applications

We're just doing the boring old topics of planar graphs and Euler's formula. Total joke—these are pretty much the most fun ever.

11.1   Chapter 11 Definitions and Notation

11.1.1   Notation

K_n: The complete graph on n vertices (just as a reminder).

$K_{n,m}$: The complete bipartite graph with n vertices in one part and m vertices in the other part (just as a reminder).

11.1.2   Definitions

planar: A graph that can be drawn in the plane (on a piece of paper, on the blackboard, etc.) without edges crossing.

face: A contiguous area of the plane bounded by edges, in a planar drawing of a planar graph.

size (of a face): The number of edges bounding the face. (Sometimes one edge appears twice on the boundary of a face, in which case it is counted twice.)

thickness (of a graph): For a graph G with n vertices, the smallest number of n-vertex planar graphs that can be stacked up (aligning the vertices) to form G.

girth: The length of a graph's smallest cycle.

11.2   Chapter 11 Facts and Theorems

Fun fact.   The website where you can play with planar graphs is http://planarity.net.

Theorem 11.3.1.   For any graph G drawn without edges crossing, the sum of the sizes of the faces is equal to $2|E(G)|$.

Theorem 11.5.1, Euler's formula.   For any connected planar graph G with faces $F(G)$, $|V(G)| - |E(G)| + |F(G)| = 2$.

Theorem 11.6.1.   If G is simple, planar, and connected and has at least three vertices, then $|E(G)| \leq 3|V(G)| - 6$.

Theorem 11.6.2.   If G is simple, planar, and connected and has at least three vertices, then $3|F(G)| \leq 2|E(G)|$.

Theorem 11.6.3.   If G is simple, planar, connected, and has no 3-cycles, then $4|F(G)| \leq 2|E(G)|$.

Theorem 11.6.4.   If G is simple, planar, connected, has no 3-cycles, and has at least three vertices, then $|E(G)| \leq 2|V(G)| - 4$.

Theorem 11.6.5.   Both K_5 and $K_{3,3}$ are nonplanar.

Theorem 11.6.6.   If G is simple, planar, and connected, then G has at least one vertex of degree no more than 5.

Theorem 11.6.7.   If G is a simple graph with at least three vertices, then $t(G) \geq \left\lceil \frac{|E(G)|}{3|V(G)|-6} \right\rceil$.

11.3   Some Straightforward Examples of Chapter 11 Ideas

An example of a planar graph.   Figure 11.1 shows a planar graph; at left is a nonplanar drawing of the graph and at right is a planar drawing of the graph.

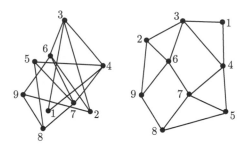

Figure 11.1. Nonplanar and planar drawings of the same graph.

An example of a nonplanar graph.   Figure 11.2 shows a nonplanar graph with a $K_{3,3}$ subgraph highlighted. One part of $K_{3,3}$ has white vertices and the other has grey vertices. (Because the graph contains $K_{3,3}$, we know it

Figure 11.2. A nonplanar graph with highlighted $K_{3,3}$ subgraph.

is nonplanar by Theorem 11.6.5.)

An example of a nonplanarity proof using Euler's formula.   You are shown a graph that looks suspiciously like a scribbly duck, and asked whether or not it is planar. A quick count shows that it has 23 vertices and 65 edges. *Fine,* you think, *it could be planar. After all,* $23 - 65 + 44 = 2$, *so it just needs 44 faces.* But somehow you feel unsettled by trying to draw all those faces. *Maybe there's a consequence of Euler's formula that will* *help? Ah, yes, Theorem 11.6.1*—and $65 \nleq 3 \cdot 23 - 6$, so we now know the scribbly duck graph cannot be planar.

11.4   More Problems for Chapter 11

Those solutions that model a formal write-up (such as one might hand in for homework) are to Problems 2, 7, and 9.

1. Compute the thickness of K_6.

2. Check out Figure 11.3 to see an image of an annulus (or washer).

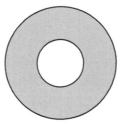

Figure 11.3. An annulus. Or a washer. Who knows?

   (a) Draw a few graphs on annuli (that's the plural of *annulus*). The rule here is that you have to cover the annulus edges with graph edges (and vertices) so that you don't have partial faces.

   (b) Try out Euler's formula on these graphs. Does it still hold? If not, does some other formula hold?

   (c) Prove your conjecture.

3. Is the complement of any star graph planar? Are *all* complements of star graphs planar? Justify your responses.

4. Can there exist a planar graph with degree sequence $(1,2,2,2,3,5,5,6)$?

5. Could the graph at left in Figure 11.4 be planar?

6. The graph at right in Figure 11.4 is definitely planar. How many faces does a planar drawing of this graph have?

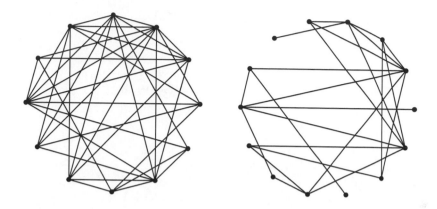

Figure 11.4. Two random graphs courtesy of *Mathematica*.

7. Prove that a connected planar graph has exactly one face if and only if it is a tree.

8. How many vertices must a 4-regular planar graph with 12 faces have?

9. Can a planar graph with 9 vertices and all faces of size 4 be k-regular for any k?

10. Compute the thickness of the nonplanar Grötzsch graph, shown in Figure 11.5.

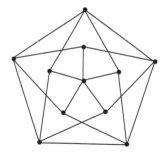

Figure 11.5. I am named after Herbert (Camillo) Grötzsch.

11.5   More Solutions for Chapter 11

1. Compute the thickness of K_6.

   First, let's see what Theorem 11.6.7 says: $t(K_6) \geq \left\lceil \frac{15}{3 \cdot 6 - 6} \right\rceil = 2$. This isn't particularly enlightening—we already knew that K_6 is nonplanar. If we can exhibit K_6 as having thickness 2 we'll be done. See Figure 11.6 for such an exhibit.

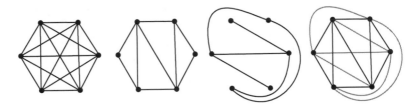

   Figure 11.6. Ordinary K_6; two planar graphs that make K_6 when glued at the vertices; K_6 drawn to highlight the two planar graphs that comprise it.

2. Check out Figure 11.3 to see an image of an annulus (or washer).

   (a) Draw a few graphs on annuli (that's the plural of *annulus*). The rule here is that you have to cover the annulus edges with graph edges (and vertices) so that you don't have partial faces.

   (b) Try out Euler's formula on these graphs. Does it still hold? If not, does some other formula hold?

   (c) Prove your conjecture.

   (a) See Figure 11.7.

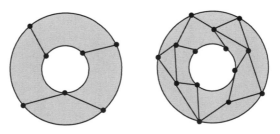

   Figure 11.7. Two sample annular graphs.

(b) For the two graphs drawn in Figure 11.7, we have $7 - 11 + 4 = 0$ and $14 - 27 + 13 = 0$. Euler's formula does not hold, but instead $|V(G)| - |E(G)| + |F(G)| = 0$.

(c) Notice that if we fill in the middle of an annulus with a face, and fill in the outside as well, we have a planar graph! The difference is that we have two extra faces. So, we have a one-to-one correspondence between annular graphs and certain planar graphs. If we have an annular graph G and associated planar graph G', then we know $|V(G)| = |V(G')|, |E(G)| = |E(G')|$, and $|F(G)| = |F(G')| - 2$.

From Euler's formula we have
$|V(G')| - |E(G')| + |F(G')| = 2$. Substituting, we get
$|V(G)| - |E(G)| + |F(G)| + 2 = 2$ or
$|V(G)| - |E(G)| + |F(G)| = 0$ as desired.

3. Is the complement of any star graph planar? Are *all* complements of star graphs planar? Justify your responses.

Let us look at a couple of the smallest star graphs and their complements, as shown in Figure 11.8. We have drawn the star graphs so that the central vertex is to the side, so that they look more like fans than stars. We notice first that these particular complements

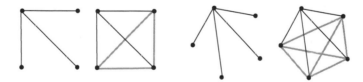

Figure 11.8. 3-spoke and 4-spoke stars shown together with their complements.

are planar (a triangle and K_4, respectively). Then we realize that if we draw K_n and use this as a base for drawing K_{n+1}, what we're adding is a star graph—one vertex with n spokes! In other words, the complement of a star graph with n spokes is K_n. Therefore, the only star graphs with planar complements are those shown above.

4. Can there exist a planar graph with degree sequence $(1,2,2,2,3,5,5,6)$?

The proposed graph has 8 vertices and total degree $1 + 2 + 2 + 2 + 3 + 5 + 5 + 6 = 26$, so it has 13 edges. Let's check Theorem 11.6.1:

$13 \le 3 \cdot 8 - 6 = 18$. So it seems possible. Can we exhibit such a graph? Yes, though it has multiple edges; see Figure 11.9 for such a graph.

Figure 11.9. A planar graph with degree sequence $(1,2,2,2,3,5,5,6)$.

5. Could the graph at left in Figure 11.4 be planar?

   Nope. It has 12 vertices and 38 edges, and $38 \nleq 3 \cdot 12 - 6 = 30$, so we have a contradiction by Theorem 11.6.1.

6. The graph at right in Figure 11.4 is definitely planar. How many faces does a planar drawing of this graph have?

   It has 14 vertices and 24 edges, so $14 - 24 + F = 2$ implies it has 12 faces.

7. Prove that a connected planar graph has exactly one face if and only if it is a tree.

   First, we know that a tree has exactly one face by inspection. So we need to prove that if a planar graph has exactly one face, then it must be a tree.

   Consider a planar graph with exactly one face. Then we have $v - e + 1 = 2$ or $v = e + 1$ or $e = v - 1$, which we know from Theorem 10.2.1 implies that the graph must be a tree.

8. How many vertices must a 4-regular planar graph with 12 faces have?

   A 4-regular planar graph with v vertices has total degree $4v$ and so has $2v$ edges. Euler's formula gives us that $v - 2v + 12 = 2$, or $v = 10$.

9. Can a planar graph with 9 vertices and all faces of size 4 be k-regular for any k?

   All faces are of size 4, so $2e = 4f$ which gives us $v - 2f + f = 2$ or $v - f = 2$ or $f = v - 2$. Having 9 vertices means there are 7 faces and 14 edges. The total degree is 28, which is not divisible by 9, so the graph cannot be regular.

10. Compute the thickness of the nonplanar Grötzsch graph, shown in
    Figure 11.5.

    First, let's see what Theorem 11.6.7 says: $t(GG) \geq \left\lceil \frac{20}{3 \cdot 11 - 6} \right\rceil = 2$.
    This isn't particularly enlightening—we already knew that the
    Grötzsch graph is nonplanar. If we can exhibit the Grötzsch graph
    as having thickness 2 we'll be done. See Figure 11.10, and notice
    that every crossing involves one of the five grey edges. Together,
    they are planar, as is the remaining black-edged graph.

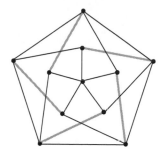

Figure 11.10. See how not very thick I am?

Chapter 12

Graph Traversals

We want to go from point A to point B in the shortest way, or to go through every vertex of a graph exactly once, or to go through every edge of a graph exactly once. So we study Dijkstra's algorithm, Hamiltonian paths and circuits, and Euler trails and circuits.

12.1   Chapter 12 Definitions and Notation

12.1.1   Definitions

Euler traversal: Visits every edge of a graph exactly once and may visit some vertices of the graph more than once.

Euler trail: An Euler traversal that is not a circuit.

Euler circuit: An Euler traversal whose starting and ending vertices are the same.

Hamilton traversal: Visits every vertex of a graph exactly once.

Hamilton path: A Hamilton traversal that is not a circuit.

Hamilton circuit: A Hamilton traversal whose starting and ending vertices are adjacent.

Hamilton cycle: A Hamilton circuit.

Traveling Salesperson's Problem: The problem of finding the *shortest* Hamilton circuit in a graph. It has so many practical applications that there is a website (http://www.tsp.gatech.edu/index.html) at which recent research and results related to TSP are compiled.

TSP: Abbreviation for Traveling Salesperson's Problem.

adjacent transposition: A permutation that switches two elements that
    are next to each other in an arrangement.

12.2   Chapter 12 Facts and Theorems

Theorem 12.3.1.
    (1) A connected graph G has an Euler circuit $\Longleftrightarrow$ every vertex of G
has even degree.
    (2) A connected graph G has an Euler traversal but not an Euler circuit $\Longleftrightarrow$ G has exactly two vertices of odd degree.

Theorem 12.4.2.   Let the simple graph G have $n \geq 3$ vertices. If the degree of every vertex is more than $\frac{n}{2}$, then G has a Hamilton circuit.

How to find the shortest path between two vertices using Dijkstra's algorithm:

1. Get ready by locating at least two colored pens, finding a graph that
   has weighted edges, and labeling the vertices with letters in one
   color.

2. Using a different color, circle the start vertex s.

3. Look at the weights of the edges incident to s. For the smallest
   weight, go along one edge corresponding to that weight and tag
   the adjacent vertex with (w_e, s), where w_e is the weight of the edge
   connecting the vertex to s. This means that the new vertex is tagged
   with the shortest distance from s and the previous vertex in the
   shortest path (in this case, it's just s).

4. Consider all the tagged vertices as a collective. For each tagged
   vertex v, compute the shortest distance from s to each of its untagged neighbors. That is, add the weight of each incident edge to
   the number in v's tag to produce a list of distances for the vertex.

5. Determine the smallest distance d among all the lists for all the
   tagged vertices.

6. Each occurrence of d corresponds to a tagged vertex connected
   to an untagged vertex. Tag each of these untagged vertices with

(d, v), where v is the label on the corresponding tagged vertex. (If an untagged vertex could have two tags (d, v_1) and (d, v_2), it does not matter which tag is selected.)

7. If there are untagged vertices, go to step 4.

12.3 Some Straightforward Examples of Chapter 12 Ideas

An example of building an Euler circuit. Figure 12.1 shows the process of building an Euler circuit from a graph. (A) First, we pick a start vertex

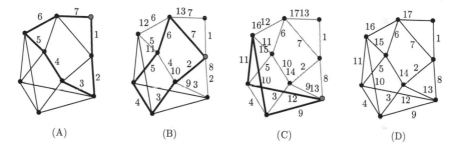

(A)   (B)   (C)   (D)

Figure 12.1. Building an Euler circuit in a graph with all even vertices.

and travel around the graph until we end where we start, labeling as we go. (B) Seeing no unused edges at our start vertex, we pick the next available vertex with unused edges and travel from there; we then incorporate this new path in our numbering. (C) We repeat the process used at (B) and exhaust the edges of the graph. (D) The original graph with complete Euler circuit.

An example of executing Dijkstra's algorithm. Figure 12.2 shows the process of finding the distance from one vertex to all others of a small graph. The tags at each step are shown in grey.

12.4 More Problems for Chapter 12

Those solutions that model a formal write-up (such as one might hand in for homework) are to Problems 3 and 7.

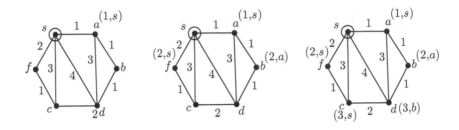

Figure 12.2. Executing Dijkstra's algorithm on a small graph.

1. An *n-prism* graph is constructed by putting one (slightly smaller) *n*-cycle C_n inside another, and adding edges to join the vertices of one C_n to the other radially. (We do need $n \geq 3$.) See Figure 12.3 for an example. Do any *n*-prism graphs have Euler circuits? What about Hamilton circuits?

Figure 12.3. I am a proud 5-prism graph.

2. List all possible orderings of ABC (how many are there?). Associate each of these orderings to a vertex of a graph. Add an edge when two orderings differ only by an adjacent transposition.

   (a) What is the degree sequence of this graph?

   (b) Does it have an Euler circuit or trail?

   (c) Does it have a Hamilton circuit or trail?

   (d) Is it planar?

   (e) What are the answers to the previous questions if we also consider the first and last letters to be adjacent?

3. Look at the graphs in Figure 10.5 on page 122. Does either have a Hamilton circuit? ... Hamilton traversal? ... Euler circuit? ... Euler traversal?

4. Again examine Figure 10.5 on page 122. For each graph, compute the shortest distance from the lower-right vertex to all other vertices. (Tip: Dijkstra is a good choice here.)

5. Do any of the graphs in Figure 12.4 have Hamilton circuits? What about Hamilton traversals?

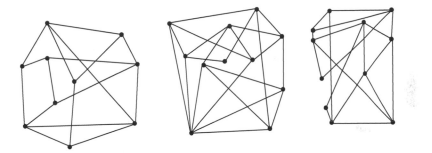

Figure 12.4. Three graphs. Yup.

6. Do any of the graphs in Figure 12.4 have Euler circuits? What about Euler traversals?

7. For which values of m, n does $K_{m,n}$ have a Hamilton circuit?

8. The towns Gesund and Reichtum are near each other in a tourism district. In each town, all but two of the intersections are 4-way stops. In Gesund, there is a 5-way stop and a "T" intersection (a 3-way stop), and in Reichtum there are two 5-way stops. Currently, there is no direct road between Gesund and Reichtum. The tourism bureau wants to build a road so that they can create and advertise a Tour of the Towns, which will take tourists down every road of Gesund and of Reichtum without repetition. What advice can you give the tourism bureau?

9. In the metropolis of Altana, the Traffic Council has decreed that cars in the flying lanes must pay twice the tolls of ground-based cars (because of the additional fuel needed for flying police). What is the cheapest way to get from point a to point b? A map showing skyways in grey and ground-roads in black is shown in Figure 12.5—those dots are toll stations where you pay for the segment you've just traveled.

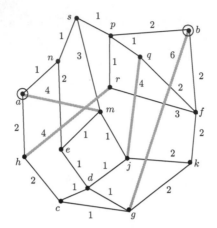

Figure 12.5. A map of Altana toll plazas.

10. Can you take one walk and cover every road in the map of Snake-
    land given in Figure 12.6 exactly once?

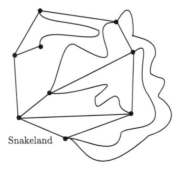

Figure 12.6. A map of Snakeland.

12.5   More Solutions for Chapter 12

1. An *n-prism* graph is constructed by putting one (slightly smaller) *n*-cycle C_n inside another, and adding edges to join the vertices of one C_n to the other radially. (We do need $n \geq 3$.) See Figure 12.3 for an example. Do any *n*-prism graphs have Euler circuits? What about Hamilton circuits?

    No Euler circuits (or even trails), by Theorem 12.3.1—all vertices of an *n*-prism have degree 3. However, as shown in Figure 12.7, every *n*-prism graph has a Hamilton circuit.

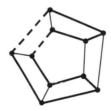

Figure 12.7. I am a general *n*-prism graph sporting a fine Hamilton circuit.

2. List all possible orderings of ABC (how many are there?). Associate each of these orderings to a vertex of a graph. Add an edge when two orderings differ only by an adjacent transposition.

    (a) What is the degree sequence of this graph?
    (b) Does it have an Euler circuit or trail?
    (c) Does it have a Hamilton circuit or trail?
    (d) Is it planar?
    (e) What are the answers to the previous questions if we also consider the first and last letters to be adjacent?

    (a) We have 6 orderings:
    ABC
    ACB
    CAB
    CBA
    BCA
    BAC. (These are written so that each differs by the next by one adjacent transposition.) Each ordering has two possible adjacent transpositions, so the degree sequence is $(2,2,2,2,2,2)$.

(b, c, d) The graph is a 6-cycle, so it is planar, and is an Euler circuit and a Hamilton circuit.

(e) If we consider the first and last letters to be adjacent, we would have three possible adjacent transpositions for each ordering; we would have degree sequence $(3,3,3,3,3,3)$; and we would add the edges ABC–CBA, ACB–BCA, and CAB–BAC. The resulting graph has a Hamilton circuit (the same as the cycle had) but no Euler circuit (all vertices are of odd degree); it is isomorphic to $K_{3,3}$, which is not planar.

3. Look at the graphs in Figure 10.5 on page 122. Does either have a Hamilton circuit? ...Hamilton traversal? ...Euler circuit? ...Euler traversal?

Each graph has at least one Hamilton circuit, as shown in Figure 12.8.

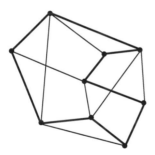 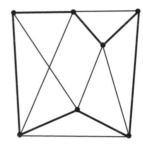

Figure 12.8. We are very Hammy.

The left-hand graph has 4 vertices of degree 3, and so has no Euler traversal by Theorem 12.3.1. The right-hand graph has two vertices of degree 3 and the rest are of degree 4, so it has an Euler trail but no Euler circuit.

4. Again examine Figure 10.5 on page 122. For each graph, compute the shortest distance from the lower-right vertex to all other vertices. (Tip: Dijkstra is a good choice here.)

We start by labeling the vertices, and then apply Dijkstra's algorithm. The results are shown in Figure 12.9.

5. Do any of the graphs in Figure 12.4 have Hamilton circuits? What about Hamilton traversals?

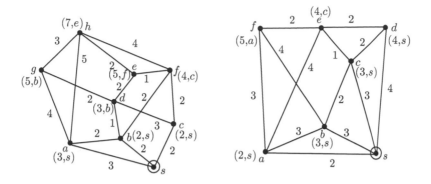

Figure 12.9. We are very Dijkstry.

All of them have Hamilton circuits, as shown in Figure 12.10.

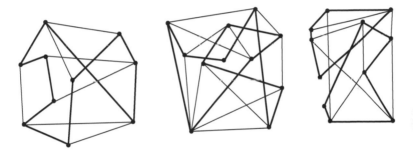

Figure 12.10. Three graphs with Hamilton circuits shown. Yup.

6. Do any of the graphs in Figure 12.4 have Euler circuits? What about Euler traversals?

The left-hand graph has more than three vertices of degree 3, so by Theorem 12.3.1 there's no chance it can have an Euler traversal. The middle graph has all vertices of even degree, so has an Euler circuit. The right-hand graph has exactly two vertices of degree 3, so has an Euler trail but not an Euler circuit.

7. For which values of m, n does $K_{m,n}$ have a Hamilton circuit?

$K_{m,n}$ has a Hamilton circuit exactly when $m = n$.

First, note that when $m = n$ we can construct a Hamilton circuit by zig-zagging as in Figure 12.11. Next, recall that in a bipartite

Figure 12.11. A Hamilton circuit in $K_{n,n}$.

graph, edges are only present between the parts, and not internal
to either part. Therefore any path in a bipartite graph must alternate between the parts. Suppose $m > n$. Our longest path (with no
vertices repeated) will only reach $n + 1$ vertices of the m vertices.
Even if $m = n + 1$, we only have a Hamilton path and not a Hamilton circuit.

8. The towns Gesund and Reichtum are near each other in a tourism
   district. In each town, all but two of the intersections are 4-way
   stops. In Gesund, there is a 5-way stop and a "T" intersection (a
   3-way stop), and in Reichtum there are two 5-way stops. Currently,
   there is no direct road between Gesund and Reichtum. The tourism
   bureau wants to build a road so that they can create and advertise
   a Tour of the Towns, which will take tourists down every road of
   Gesund and of Reichtum without repetition. What advice can you
   give the tourism bureau?

   Build *two* roads, each connecting one of Reichtum's 5-way stops
   with one of Gesund's odd-way stops. Then every intersection will
   have even degree and by Theorem 12.3.1, an Euler circuit will exist.

9. In the metropolis of Altana, the Traffic Council has decreed that
   cars in the flying lanes must pay twice the tolls of ground-based
   cars (because of the additional fuel needed for flying police). What
   is the cheapest way to get from point a to point b? A map showing skyways in grey and ground-roads in black is shown in Figure
   12.5—those dots are toll stations where you pay for the segment
   you've just traveled.

   Figure 12.12 shows the result of running Dijkstra's algorithm on
   the Altana map; the cheapest toll route costs \$5 and goes through n,
   s, and p.

10. Can you take one walk and cover every road in the map of Snake-
    land given in Figure 12.6 exactly once?

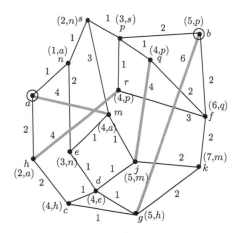

Figure 12.12. A map of Altana toll plazas.

By Theorem 12.3.1, no, because there are two vertices of degree 3.

Chapter 13

Graph Coloring

We focus on coloring, as with markers or crayons, the vertices or edges of a graph. It is preferred to use the least number of colors needed for a proper coloring.

13.1   Chapter 13 Definitions and Notation

13.1.1   Notation

$\chi(G)$: The chromatic number of graph G.

$\chi'(G)$: The chromatic index of graph G.

13.1.2   Definitions

coloring: An assignment of colors to a graph G's vertices (or edges); technically, it is a function $c : V(G) \rightarrow C$ (or $c : E(G) \rightarrow C$), where C is a set of colors.

proper vertex coloring: Each vertex of G is assigned a color such that no two adjacent vertices have the same color.

proper edge coloring: Each edge of G is assigned a color such that no two edges incident to the same vertex have the same color.

chromatic number: The smallest number of colors needed to properly color the vertices of G; we denote it by $\chi(G)$.

k-colorable: Can be properly colored using k vertex colors.

k-chromatic: Can be properly colored using *no fewer than* k vertex colors. G is always $\chi(G)$-chromatic.

chromatic index: The smallest number of colors needed to properly color the edges of G; we denote it by $\chi'(G)$.

k-edge-colorable: Can be properly colored using *k* edge colors.

k-edge-chromatic: Can be properly colored using *no fewer than k* edge colors. *G* is always $\chi'(G)$-chromatic.

13.2   Chapter 13 Facts and Theorems

Lower bounds for coloring:

* 🐦 A graph containing *n* mutually adjacent vertices (that is, a copy of K_n) will need at least *n* vertex colors. So if *G* contains K_n, then $\chi(G) \geq n$. (The converse is not true.)

* 🐦 If the maximum degree of a vertex in *G* is $\Delta(G)$, then $\chi'(G) \geq \Delta(G)$ because there are $\Delta(G)$ edges that are incident at some vertex.

Theorem 13.3.5.   Let *G* be a simple graph with largest degree $\Delta(G)$. Then $\chi(G) \leq \Delta(G) + 1$.

Theorem 13.3.6.   Let *G* be a simple graph with largest degree $\Delta(G)$. Then $\chi'(G) \leq 2\Delta(G) - 1$.

Theorem 13.5.1.   For *n* even, $\chi'(K_n) = n - 1$, and for *n* odd, $\chi'(K_n) = n$.

Theorem 13.5.2.   A graph *G* is bipartite if and only if is 2-vertex colorable.

Theorem 13.5.3.   A graph *G* is bipartite $\iff$ *G* has no odd cycles.

Theorem 13.5.4.   For bipartite *G*, $\chi'(G) = \Delta(G)$.

Theorem 13.5.5.   Every simple planar graph can be vertex-colored with at most six colors.

A greedy algorithm for coloring vertices (or edges) of a graph:

1. Order the vertices (or edges) of the graph as $v_1, \ldots, v_n$ (or as $e_1, \ldots, e_n$).

2. Make a list of colors, namely color 1, color 2, $\ldots$, color *n*.

3. Color the first vertex (or edge) with color 1.

4. Consider the next vertex (or edge) in the list, and give it the lowest-numbered color that is not already in use on one of the vertex's neighbors (or one of the edges incident to this edge).

5. If all of the vertices (or edges) are colored, be done. If not, go to step 4.

Fact. A greedy algorithm for coloring does not always give an optimal coloring (and can sometimes give an awful coloring). However, most of the time a greedy algorithm produces a pretty decent coloring, so sometimes greedy algorithms are used in practice.

Matchings and edge colorings. In a properly edge-colored graph, the edges of a single color form a matching. Thus, a proper edge coloring is a union of matchings.

13.3 Some Straightforward Examples of Chapter 13 Ideas

An example of chromatic number and chromatic index computation. We will compute the chromatic number and chromatic index of the 5-prism graph G pictured in Figure 12.3 on page 142. G contains an odd cycle, so $\chi(G) \geq 3$, and we exhibit a 3-vertex coloring of G in Figure 13.1, so $\chi(G) \leq 3$. Therefore $\chi(G) = 3$.
G has a vertex of degree 3, so $\chi'(G) \geq 3$, and we exhibit a 3-edge coloring of G in Figure 13.1, so $\chi'(G) \leq 3$. Therefore $\chi'(G) = 3$.

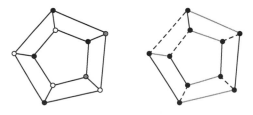

Figure 13.1. A proper vertex coloring and a proper edge coloring of the 5-prism graph.

Actually, we can compute the chromatic number and chromatic index of the n-prism graph Pr_n.

In the case that n is even, $\chi(Pr_n) = 2$; we can alternate colors on the outer cycle, and use the same kind of coloring on the inner cycle but with the colors reversed. In the case that n is odd, $\chi(Pr_n) = 3$; we can use the same scheme as in Figure 13.1.

When n is even, $\chi'(Pr_n) = 3$; we can alternate colors on the outer cycle, and use the same coloring on the inner cycle, and use the third color on all of the struts. In the case that n is odd, $\chi'(Pr_n) = 3$; we can use the same scheme as in Figure 13.1.

An example of a greedy algorithm gone bad.   Every tree T has $\chi(T) = 2$. (This is because every tree is bipartite.) Construct a tree and order the vertices of that tree such that when you color the vertices using a greedy/parsimonious algorithm, you need at least three colors.

We will do even better—we will show that a path graph with 4 vertices has this property (and therefore any tree containing a path of length 4 has this property). When the vertices are ordered as in Figure 13.2, our algorithm proceeds as follows:

Figure 13.2. A pathetic little path graph.

Vertex 1 is assigned color *black*.

Vertex 2 is not adjacent to any colored vertices, so it can also be assigned color *black*.

Vertex 3 is adjacent to a vertex with color *black*, so we must use color *white*.

Vertex 4 is adjacent to vertices colored *black* and *white*, so we are forced to use color *grey*.

Example 13.3.9 rewritten.   Bad traffic is super-annoying, especially at traffic lights. Let us try to find a way to allow as many lanes to have green lights as possible at the same time, while also keeping drivers from colliding with each other. We will use vertex coloring, and so we must first create a graph to color.

Proper graph coloring prohibits adjacent vertices from having the same color, so here we want to prohibit cars that might collide from traveling at the same time. Thus we will let the lanes be vertices, and we will let potential lane-occupant collisions be edges. As shown in Figure 13.3, we can number the lanes and use these numbers as vertex labels and see which paths of travel intersect to find edges in the corresponding graph.

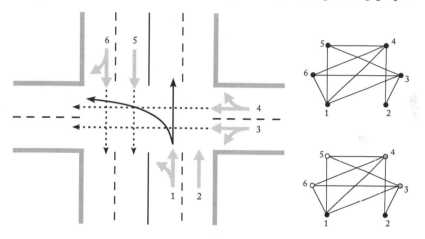

Figure 13.3. Cars traveling in lane 1 intersect paths of travel from lanes 3, 4, 5, and 6 (left). The graph corresponding to this intersection along with a proper vertex coloring (right).

Sometimes an intersection is sufficiently busy that it's very difficult to turn left. In such a situation, we have a protected left turn, indicated by a green arrow on the traffic light. When such an option exists, we consider that traffic path as a separate vertex.

13.4   More Problems for Chapter 13

Those solutions that model a formal write-up (such as one might hand in for homework) are to Problems 2, 4, and 5.

1. Find the chromatic number and chromatic index of the graph shown in Figure 10.1 on page 119.

2. Prove that if $\chi(G) \geq 3$, then G must contain an odd cycle.

3. Find the chromatic number and chromatic index of each graph shown in Figure 10.4 on page 121.

4. Find the chromatic number and chromatic index of the graph shown in Figure 11.5 on page 133.

5. Let G be a planar graph with smallest cycle length (girth) 6. Let $v_G = |V(G)|, e_G = |E(G)|$, and $f_G = |F(G)|$.

    (a) Develop an inequality that relates f_G to e_G.

    (b) Use this to show that $2e_G \leq 3v_G - 6$.

    (c) Show that G must have a vertex of degree less than 3.

    (d) Prove that $\chi(G) \leq 3$. (Hint: use induction.)

6. Without doing any actual coloring, give quick lower and upper bounds for the chromatic number and chromatic index of the graph shown in Figure 13.4.

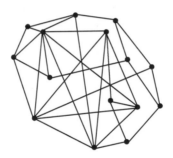

Figure 13.4. A semi-mysterious graph.

7. Find the chromatic number and chromatic index of the Snakeland map graph shown in Figure 12.6 on page 144.

8. Extend the idea in our Example of a Greedy Algorithm Gone Bad (in Section 13.3): Given any $k \geq 3$, construct a tree and order the vertices of that tree such that when you color the vertices using a greedy/parsimonious algorithm, you need at least k colors.

9. Find the chromatic number of each graph shown in Figure 12.4 on page 143.

10. Find the chromatic index of each graph shown in Figure 12.4 on page 143.

13.5   More Solutions for Chapter 13

1. Find the chromatic number and chromatic index of the graph shown in Figure 10.1 on page 119.

   Let the pictured graph be G. $\chi(G) = 3$; G contains K_3, so $\chi(G) \geq 3$, and we exhibit a 3-vertex coloring of G in Figure 13.5.
   $\chi'(G) = 4$; G has a vertex of degree 4, so $\chi'(G) \geq 4$, and we exhibit a 4-edge coloring of G in Figure 13.5.

   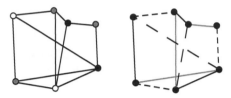

   Figure 13.5. A proper vertex coloring and a proper edge coloring of the graph in question.

2. Prove that if $\chi(G) \geq 3$, then G must contain an odd cycle.

   We proceed by contradiction. Suppose that G has no odd cycles; then by Theorem 13.5.3, G is bipartite and then by Theorem 13.5.2, G is 2-vertex colorable. Contradiction! We know $\chi(G) \geq 3$; therefore, G must contain an odd cycle.

3. Find the chromatic number and chromatic index of each graph shown in Figure 10.4 on page 121.

   Let the left-hand graph be G and the right-hand graph be H.
   $\chi(G) = 3$; G contains K_3, so $\chi(G) \geq 3$, and we exhibit a 3-vertex coloring of G in Figure 13.6.
   $\chi'(G) = 6$; G has a vertex of degree 6, so $\chi'(G) \geq 6$, and we exhibit a 6-edge coloring of G in Figure 13.6.
   $\chi(H) = 3$; H contains K_3, so $\chi(H) \geq 3$, and we exhibit a 3-vertex coloring of H in Figure 13.6.
   $\chi'(H) = 4$; H has a vertex of degree 4, so $\chi'(H) \geq 4$, and we exhibit a 4-edge coloring of H in Figure 13.6.

4. Find the chromatic number and chromatic index of the graph shown in Figure 11.5 on page 133.

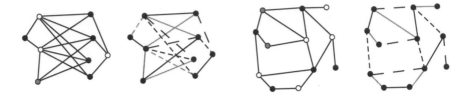

Figure 13.6. Proper vertex coloring and proper edge coloring of each graph in question.

Let the pictured graph be GG. $\chi(GG) = 4$. We know that the outer 5-cycle requires 3 colors, so let us suppose that there exists a 3-coloring of GG. Because GG has 5-fold rotational symmetry, we can 3-color that cycle as we please. This coloring forces the colors on the spokes of the inner star, and all three colors are present there. The middle vertex is adjacent to all of them, so it requires a fourth color. To see the 4-coloring, refer to Figure 13.7.

$\chi'(GG) = 5$; GG has a vertex of degree 5, so $\chi'(GG) \geq 5$, and we exhibit a 5-edge coloring of GG in Figure 13.7.

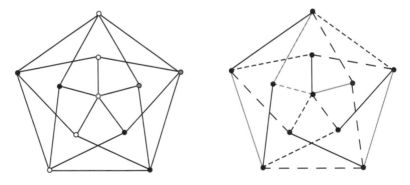

Figure 13.7. A proper vertex coloring and a proper edge coloring of the Grötzsch graph.

5. Let G be a planar graph with smallest cycle length (girth) 6. Let $v_G = |V(G)|, e_G = |E(G)|$, and $f_G = |F(G)|$.

   (a) Develop an inequality that relates f_G to e_G.

   (b) Use this to show that $2e_G \leq 3v_G - 6$.

   (c) Show that G must have a vertex of degree less than 3.

   (d) Prove that $\chi(G) \leq 3$. (Hint: use induction.)

(a) First note that because the graph is planar, we can use Euler's formula and its consequences. Because the smallest cycle length is 6, then every face has size at least 6. The sum of the sizes of the faces is equal to $2e_G$. If all the faces were size 6, we would have $6f_G = 2e_G$, but some faces may be larger, and so we have $6f_G \leq 2e_G$.

(b) Euler's formula says that $v_G - e_G + f_G = 2$ or $f_G = 2 - v_G + e_G$. Plugging this into $6f_G \leq 2e_G$ gives us
$6(2 - v_G + e_G) = 12 - 6v_G + 6e_G \leq 2e_G$ or $6 - 3v_G + 3e_G \leq e_G$ or $2e_G \leq 3v_G - 6$.

(c) We proceed by contradiction. Suppose that G has only vertices of degree 3 or more. Then the total degree of G is at least $3v_G$. By the handshaking lemma, the total degree of G is $2e_G$. Therefore $2e_G \geq 3v_G$. However, $2e_G \leq 3v_G - 6$, so we have $3v_G \leq 2e_G \leq 3v_G - 6$ or $3v_G \leq 3v_G - 6$ or $0 \leq -6$ which is a contradiction— therefore G must have a vertex of degree less than 3.

(d) Now we proceed by induction on the number of vertices of G. As a base case we take $v_G = 6$ so G has at least one cycle. We know this can be colored with only two colors, so $\chi(G) \leq 3$. Assume (for an inductive hypothesis) that if G has $6 \leq n \leq k$ vertices, then $\chi(G) \leq 3$. Consider G with $k + 1$ vertices and girth at least 6. We know that G must have a vertex y of degree less than 3. Remove this vertex to form H. The inductive hypothesis applies to H because either H still has girth at least 6 or H has no cycles (in which case it is 2-vertex colorable). So, color H with at most 3 colors. The vertex y has at most two neighbors, so it can be given a third color—we restore y to the graph, colored with this third color, and have G properly vertex colored using at most 3 colors. *Finis.*

6. Without doing any actual coloring, give quick lower and upper bounds for the chromatic number and chromatic index of the graph shown in Figure 13.4.

The graph contains a K_5 and has highest degree 8, so $5 \leq \chi(G) \leq 9$. The graph has highest degree 8, so $8 \leq \chi'(G) \leq 15$.

7. Find the chromatic number and chromatic index of the Snakeland map graph shown in Figure 12.6 on page 144.

Let the Snakeland map graph be $ssss$. $\chi(ssss) = 3$; $ssss$ contains K_3, so $\chi(ssss) \geq 3$, and we exhibit a 3-vertex coloring of $ssss$ in Figure 13.8.

$\chi'(ssss) = 4$; *ssss* has a vertex of degree 4, so $\chi'(ssss) \geq 4$, and we exhibit a 4-edge coloring of *ssss* in Figure 13.8.

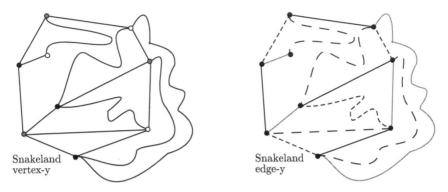

Figure 13.8. A proper vertex coloring and a proper edge coloring of the Snakeland map graph.

8. Extend the idea in our Example of a Greedy Algorithm Gone Bad (in Section 13.3): Given any $k \geq 3$, construct a tree and order the vertices of that tree such that when you color the vertices using a greedy/parsimonious algorithm, you need at least k colors.

The start of a construction (or a construction for $k = 5$) is shown in Figure 13.9. The general idea is to take ordered paths of lengths

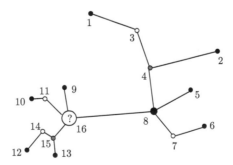

Figure 13.9. A weird start of a tree.

1, 2, and 4 to produce vertices colored *black*, *white*, and *grey*, and add another vertex adjacent to those three vertices that must be colored *BLACK*. Use this 8-vertex gadget along with ordered paths of lengths 1, 2, and 4 to produce vertices that must be colored *black*,

white, *grey*, and *BLACK*; add another vertex adjacent to those four
vertices (it must be colored *huh?*). This process of using paths and
increasingly large gadgets can be repeated as many times as necessary to force a vertex in a kth color.

9. Find the chromatic number of each graph shown in Figure 12.4 on
   page 143.

   Let the left-hand graph be G, the middle graph be Q, and the right-
   hand graph be H. $\chi(G) = 3$; G contains K_3, so $\chi(G) \geq 3$, and we
   exhibit a 3-vertex coloring of G in Figure 13.10.
   $\chi(Q) = 3$; Q contains K_4, so $\chi(Q) \geq 4$, and we exhibit a 4-vertex
   coloring of Q in Figure 13.10.
   $\chi(H) = 3$; H contains K_3, so $\chi(H) \geq 3$, and we exhibit a 3-vertex
   coloring of H in Figure 13.10.

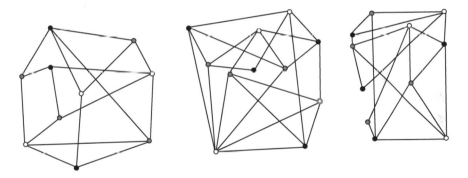

Figure 13.10. A proper vertex coloring of each graph in question.

10. Find the chromatic index of each graph shown in Figure 12.4 on
    page 143.

    Let the left-hand graph be G, the middle graph be Q, and the right-
    hand graph be H. $\chi'(G) = 4$; G has a vertex of degree 4, so $\chi'(G) \geq$
    4, and we exhibit a 4-edge coloring of G in Figure 13.11.
    $\chi'(Q) = 6$; Q has a vertex of degree 6, so $\chi'(Q) \geq 6$, and we exhibit
    a 6-edge coloring of Q in Figure 13.11.
    $\chi'(H) = 4$; H has a vertex of degree 4, so $\chi'(H) \geq 4$, and we ex-
    hibit a 4-edge coloring of H in Figure 13.11.

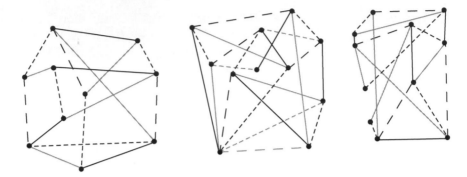

Figure 13.11. A proper edge coloring of each graph in question.

Part IV

Other Material

Chapter 14

Probability and Expectation

In our study of discrete probability, we look only at situations with a finite number of possibilities. In this context, we look at conditional probability, random variables, and expected value, with a nod to the supremely powerful probabilistic method.

14.1 Chapter 14 Definitions and Notation

14.1.1 Notation

$P(E)$: The probability that event E occurs.

$X(s)$: The value of random variable X for state s.

$X = k$: The subset of $s \in S$ such that $X(s) = k$.

$\mathbb{E}[X]$: The expected value of a random variable X, defined as $\mathbb{E}[X] = \sum_{k \in N} kP(X = k)$.

$P(E_1|E_2)$: The conditional probability that event E_1 happens, given that E_2 definitely occurs. $P(E_1|E_2) = \frac{P(E_1 \text{ and } E_2)}{P(E_2)}$.

$X_1 + X_2$: The sum of two random variables is defined pointwise as $(X_1 + X_2)(s) = X_1(s) + X_2(s)$.

14.1.2 Definitions

probability: The likelihood that some given event will occur.

state: A configuration that a system might have.

state space: The set of different possible states.

sample space: A state space.

event: A collection of states; a subset of the state space.

exclusive: Two events that cannot happen at the same time. States are
          always exclusive; some events are exclusive and some are not.

probability axioms: Conditions that all probability functions must satisfy
          in order to actually measure probability, as follows.
          Let $P : \mathscr{P}(S) \to [0, 1]$ be a function from all subsets of a state space
          to the unit interval $\{x \mid 0 \leq x \leq 1\}$.

   ❧ $P(S) = 1$, that is, *some* event definitely occurs. In other
     words, the probability that the system is in *some* state is one.

   ❧ $P(\emptyset) = 0$, or, there is no possibility that nothing happens. In
     other words, the probability that the system is not in any state
     is zero.

   ❧ For any element $s \in S$, $P(s) \geq 0$ and likewise for any subset
     (event) $E \subset S$, $P(E) \geq 0$. That is, it does not make sense to
     have a negative probability.

   ❧ Whenever two events E_1, E_2 are exclusive, meaning they can-
     not happen at the same time, $P(E_1 \text{ or } E_2) = P(E_1) + P(E_2)$.
     Note that any two states are exclusive, so their probabilities
     always have this property.

random variable: A function $X : S \to N$ from a state space S to a finite set
          of real numbers N.

probability distribution: A display of all possible values for a random
          variable X with the corresponding probabilities.

expected value: Applied to a random variable X, denoted $\mathbb{E}[X]$, and de-
          fined as $\mathbb{E}[X] = \sum_{k \in N} kP(X = k)$.

conditional probability: The probability that event E_1 happens, given
          that E_2 definitely occurs. Denoted $P(E_1 | E_2)$ and measured by
          $\frac{P(E_1 \text{ and } E_2)}{P(E_2)}$.

independent: Two events A and B are independent if $P(A|B) = P(A)$ and
          $P(B|A) = P(B)$. In other words, if the probability that A happens
          is the same whether B happens or not, and if the probability that B
          happens is the same whether A happens or not, then events A and B
          are independent of each other.

14.2   Chapter 14 Facts and Theorems

Lemma 14.3.10. $\displaystyle\sum_{k\in N} kP(X=k) = \sum_{s\in S} X(s)P(s).$

Interesting fact. If A and B are independent, then $P(A|B) = \frac{P(A \text{ and } B)}{P(B)} = P(A)$, so that $P(A \text{ and } B) = P(A)P(B)$, and likewise, $P(B|A) = \frac{P(B \text{ and } A)}{P(A)} = P(B)$, with the same result. This gives us a tool for computing probabilities of multiple independent events as well as a criterion for determining whether events are independent. In fact, $P(A|B) = P(A) \iff P(B|A) = P(B)$; this means we only need to check one of the two conditions in practice.

Probability PIE. In its simplest form for sets A, B, PIE says that $|A \cup B| = |A| + |B| - |A \cap B|$. The same principle can apply to probabilities of events E_1, E_2, so that $P(E_1 \text{ or } E_2) = P(E_1) + P(E_2) - P(E_1 \text{ and } E_2)$. (And, of course, PIE generalizes to probabilities of many events in the same way that PIE generalized for sets.)

Independent vs. exclusive. We have earlier noted that two events being independent means that one event has no bearing on whether the other happens or not. We know that two events being exclusive means they cannot happen at the same time. So they are dependent on each other! If E_1 and E_2 are exclusive, when E_1 happens, then $P(E_1) = 1$ and this implies $P(E_2) = 0$. That is, $P(E_1 \cap E_2) = 0$. In other words, independence and exclusivity are in some sense opposite ends of a spectrum.

Theorem 14.7.1. For a bunch of random variables $X_1, X_2, \ldots, X_n$ on a state space S, with $X_1 + X_2 : S \to N$ defined by $(X_1 + X_2)(s) = X_1(s) + X_2(s)$, we have that $\mathbb{E}_S[X_1 + X_2] = \mathbb{E}_S[X_1] + \mathbb{E}_S[X_2]$, and, in fact, $\mathbb{E}_S[\sum_{j=1}^n X_j] = \sum_{j=1}^n \mathbb{E}_S[X_j]$. (Notice that we did not require any of the events involved in these random variables to be independent or exclusive! That makes the result somewhat surprising.)

The probabilistic method. This is an approach to proofs whereby one proves existence by showing that the probability of the desired object occurring is greater than zero. It's complicated enough to merit entire books.

14.3   Some Straightforward Examples of Chapter 14 Ideas

Examples 14.3.4 and 14.3.6 rewritten.   Consider four ducks that live together: one is white, one is white with grey spots, one is grey with white spots, and one is black with white spots. Together, all subsets of these four ducks form our state space C_d.

We define a random variable $W : C_d \to \{0,1,2,3,4\}$ as $W(d) = 1$ if the duck in question has some white, and $W(d) = 0$ if the duck has no white on it. For a subset A of the ducks, $W(A) = |A|$ because every duck has some white on it.

Similarly, define $WH : C_d \to \{0,1\}$ as $WH(d) = 1$ if the duck in question is all white, and $WH(d) = 0$ if the duck is not all white. Then for a subset A of the ducks, $WH(A) = 1$ if the all-white duck is in A, and $WH(A) = 0$ if the all-white duck is not in A.

Let us determine the probability distribution of the random variable W. Suppose it is equally likely that any subset of the ducks appears. There are $2^4 = 16$ subsets of the four ducks, and of those, one has zero ducks, four contain one duck, six contain two ducks, four contain three ducks, and one contains all four ducks. The resulting distribution is as follows:

k	$(P(W = k))$
0	$\frac{1}{16}$
1	$\frac{4}{16}$
2	$\frac{6}{16}$
3	$\frac{4}{16}$
4	$\frac{1}{16}$

To obtain the probability distribution for the random variable WH, we may look at the duck subsets and discover that there are eight subsets that include the all-white duck and eight subsets without the all-white duck. Thus, $P(WH = 0) = \frac{1}{2} = P(WH = 1)$, which we could show as in Figure 14.1.

Let us compute the expected value for W and for WH. By definition, $\mathbb{E}[W] = 0P(W = 0) + 1P(W = 1) + 2P(W = 2) + 3P(W = 3) + 4P(W = 4)$. We use the probability distribution to obtain $\mathbb{E}[W] = 0 \cdot \frac{1}{16} + 1 \cdot \frac{4}{16} + 2 \cdot \frac{6}{16} + 3 \cdot \frac{4}{16} + 4 \cdot \frac{1}{16} = 2$ white ducks. The practical interpretation is that we expect to see two ducks.

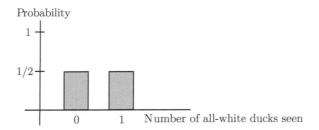

Figure 14.1. The uninteresting probability distribution for the random variable *WH*.

Similarly, $\mathbb{E}[WH] = 0 \cdot \frac{1}{2} + 1 \cdot \frac{1}{2} = \frac{1}{2}$ white duck. That is, half the time we expect to see the white duck.

Example 14.5.3 rewritten. We have a group of friends; $\frac{2}{3}$ of them own cats. A local veterinary office sends out reminder postcards; $\frac{3}{4}$ of the cat owners and $\frac{1}{5}$ of the friends who do not own cats receive these postcards.

Question 1: What fraction of the friends receive postcards? Formally, our state space is the set of friends. There are two random variables that naturally arise from this situation,

$$C(\text{friend } F) = \begin{cases} 1 & F \text{ owns a cat,} \\ 0 & F \text{ owns no cats;} \end{cases}$$

$$R(\text{friend } F) = \begin{cases} 1 & F \text{ receives a postcard,} \\ 0 & F \text{ receives no postcard.} \end{cases}$$

We want to know $P(R = 1)$, the probability that a friend receives a postcard. We can view this as (the probability that a friend receives a postcard and owns a cat) plus (the probability that a friend receives a postcard and owns no cats).

We know that $P(C = 1) = \frac{2}{3}$. The fact that $\frac{3}{4}$ of our cat-owning friends receive postcards needs a conditional probability interpretation: it is $P(R = 1 | C = 1)$. Now, (the probability that a friend receives a postcard and owns a cat) $= P(R = 1 | C = 1)P(C = 1) = \frac{1}{2}$. Similarly, we have (the probability that a friend receives a postcard and owns no cats) $= P(R = 1 | C = 0)P(C = 0) = \frac{1}{15}$. Finally, $P(R = 1) = \frac{1}{2} + \frac{1}{15} = \frac{17}{30}$.

Question 2: If a postcard arrives, what is the probability that the recipient owns a cat? We want to measure the probability that a friend owns a cat given that s/he has received a postcard, or $P(C = 1 | R = 1)$. By

definition, this is $\frac{P(C=1 \text{ and } R=1)}{P(R=1)}$. We also know that $P(C = 1 \text{ and } R =$
$1) = P(R = 1 \text{ and } C = 1) = P(R = 1|C = 1)P(C = 1) = \frac{1}{2}$. Thus,
$P(C = 1|R = 1) = \frac{1/2}{17/30} = \frac{15}{17}$.

Question 3: Are the two events *receiving a postcard* and *owning a cat* independent?

First approach: Check to see whether $P(C = 1 \text{ and } R = 1) =$
$P(C = 1)P(R = 1)$. We know $\frac{1}{2} \neq (\frac{2}{3})(\frac{17}{30}) = \frac{17}{45}$, so these events are
not independent.

Second approach: Note that $P(C = 1) \neq P(C = 1|R = 1)$ (because $\frac{1}{2} \neq \frac{15}{17}$)
and that $P(R = 1) \neq P(R = 1|C = 1)$ (because $\frac{17}{30} \neq \frac{3}{4}$), so these events are
not independent.

An example of linearity of expectation. There are generally a dozen cat toys in rotation at any one time in the author's house. Every week when the vacuuming is done, these are rounded up and put in a pile, and every week the cats redistribute toys through the house. On any given day, what is the expected number of toys removed from the pile?

The random variable T describes the number of toys in the pile. We will write $T = \sum_{i=1}^{12} T_i$, where

$$T_i(\text{day } d) = \begin{cases} 1 & \text{toy } T_i \text{ is grabbed,} \\ 0 & \text{toy } T_i \text{ is in the pile} \end{cases}$$

because it is easier to compute with the T_i.

$P(T_i = 1) = \frac{1}{7}$ because it is equally likely that a cat will first grab a toy on any day, so $\mathbb{E}[T_i] = 1P(T_i = 1) + 0P(T_i = 0) = \frac{1}{7}$.

In turn, by Theorem 14.7.1, $\mathbb{E}[T] = \sum_{i=1}^{12} \mathbb{E}[T_i] = \sum_{k=1}^{12} \frac{1}{7} = \frac{12}{7}$. Thus, we expect there to be almost 2 toys removed from the pile on any given day.

14.4   More Problems for Chapter 14

Those solutions that model a formal write-up (such as one might hand in for homework) are to Problems 2, 5, and 6.

1. In Lucy Worsley's *If Walls Could Talk: An Intimate History of the Home*, she says, "The medieval death rate was one in every fifty pregnancies. Considering that it wasn't unusual for a woman to give birth a dozen times, the odds quickly mounted up for repro-ductive wives."

So ... what are these odds? Compute the probability of dying while pregnant for each of 1, 4, 6, and 12 pregnancies. What is the probability of dying during some one of 12 theoretical medieval pregnancies?

2. Suppose you have a box of colored pens (fuchsia, cinnamon, tangerine, gold, lime, forest, teal, cobalt, plum) and three pencils (mechanical, yellow No. 2, printed with cupcakes).

   (a) Describe the state space of grabbing a pen and a pencil. What is the probability of each individual state?

   (b) What is the probability of grabbing a pen whose color begins with "f" and a mechanical pencil?

   (c) What is the probability of grabbing a pen whose color is greenish and a non-mechanical pencil?

   (d) What is the probability of (grabbing a pen whose color begins with "f" and a non-mechanical pencil) or (grabbing a pen whose color is greenish and a non-mechanical pencil)?

3. A computer lab has 20 computers in it. On any given day, the probability that a given computer is not working is p. How many computers do you expect will be functioning when you enter the lab today? Answer the question for $p = .001, p = .05$.

4. Chips of the World come in lots of flavors. In the sale bin are
2 bags of bacon ranch pita chips,
1 bag of salt and vinegar potato chips,
3 bags of hot-sauce cheese corn chips,
5 bags of crab potato chips, and
2 bags of peppercorn salsa pita chips.
If you close your eyes and grab three bags of chips (one at a time, so you know you have three), what is the probability you will get
...

   🐤 ... all three bags of potato chips?

   🐤 ... exactly 2 bags of spicy chips?

   🐤 ... at least one bag of pita chips?

How many bags of corn chips do you expect to find in your three bags?

5. Shoes 'R' Us has a lot of different kinds of shoes in their display case, one of each kind they sell. A shoe can be brown, black, silver, or green; it can be a low shoe, a boot, or an athletic shoe; and, it can have laces or be a slip-on.

   (a) How many different kinds of shoes does Shoes 'R' Us have in its display case?

   (b) What is the probability that a Shoes 'R' Us display shoe is brown and slip-on?

   (c) What is the probability that a Shoes 'R' Us display shoe is silver or a boot?

   (d) Given that a Shoes 'R' Us display shoe is silver, what is the probability that it is a boot?

   (e) Given that a Shoes 'R' Us display shoe is a green athletic shoe, what is the probability that it has laces?

   (f) Are the properties *silver* and *boot* independent?

6. Consider a deck of cards that is standard, except for having six suits—the two additional suits are stars and squids. (This deck exists: it is the Blue Sea Deck.) Draw a card.

   (a) What is the probability that the card is a queen or a squid?

   (b) What is the expected value of the number on the card? (Here, Ace = 1, King = 13.)

7. The game *Elder Sign* has unusual dice. There are six 6-sided green dice, each of which has three sides showing magnifying glasses, one side with a tentacle, one side with a skull, and one side with a scroll. There is also a 6-sided yellow die with four sides showing magnifying glasses, one side with a skull, and one side with a scroll. Finally, there is a 6-sided red die with three sides showing magnifying glasses, one side with a Wild sign, one side with a skull, and one side with a scroll.

   (a) If you roll the six green dice, what is the expected number of magnifying glasses you'll see?

   (b) If you roll all seven of the dice, what is the probability that you will roll exactly one skull?

   (c) If you roll all seven of the dice, what is the probability that you will roll at least one scroll?

8. The game of Qwirkle uses a bag of tiles. Each black tile has a shape on it (circle, diamond, square, crisscross, starburst, clover) that is colored (red, orange, yellow, green, blue, purple). There are three copies of each kind of tile.

   (a) How many tiles are in a Qwirkle bag?

   (b) What is the probability that a tile drawn is red?

   (c) What is the probability that a tile drawn is a sunburst?

   (d) What is the probability that a tile drawn is a red sunburst?

   (e) What is the probability that a tile drawn is red or a sunburst?

   (f) Is red-ness independent of sunburst-ness?

9. Another Qwirkle qwestion: Pull two tiles from the bag.

   (a) What is the probability that both are blue?

   (b) What is the probability that the *second* tile is blue?

   (c) What is the probability that at least one tile is blue?

10. What's the expected number of fixed points (items that do not move) in a permutation of n items?

14.5   More Solutions for Chapter 14

1. In Lucy Worsley's *If Walls Could Talk: An Intimate History of the Home*, she says, "The medieval death rate was one in every fifty pregnancies. Considering that it wasn't unusual for a woman to give birth a dozen times, the odds quickly mounted up for reproductive wives."

   So ... what are these odds? Compute the probability of dying while pregnant for each of 1, 4, 6, and 12 pregnancies. What is the probability of dying during some one of 12 theoretical medieval pregnancies?

   One pregnancy: $\frac{1}{50} = .02$.
   Four pregnancies: If you make it to the fourth pregnancy, that means you didn't die in the first three pregnancies, so
   $\left(\frac{49}{50}\right)^3 \left(\frac{1}{50}\right) = .0188$.
   Six pregnancies: $\left(\frac{49}{50}\right)^5 \left(\frac{1}{50}\right) = .018$.
   A dozen pregnancies: $\left(\frac{49}{50}\right)^{11} \left(\frac{1}{50}\right) = .016$.
   But dying during the first pregnancy is exclusive of dying during the second (... or twelfth) pregnancy, so in order to compute the probability of dying during *some* pregnancy of twelve, we need to
   compute $\displaystyle\sum_{j=1}^{12} \left(\frac{49}{50}\right)^{j-1} \left(\frac{1}{50}\right) = .2153$.

2. Suppose you have a box of colored pens (fuchsia, cinnamon, tangerine, gold, lime, forest, teal, cobalt, plum) and three pencils (mechanical, yellow No. 2, printed with cupcakes).

   (a) Describe the state space of grabbing a pen and a pencil. What is the probability of each individual state?

   (b) What is the probability of grabbing a pen whose color begins with "f" and a mechanical pencil?

   (c) What is the probability of grabbing a pen whose color is greenish and a non-mechanical pencil?

   (d) What is the probability of (grabbing a pen whose color begins with "f" and a non-mechanical pencil) or (grabbing a pen whose color is greenish and a non-mechanical pencil)?

   (a) The state space is {fuchsia/mechanical, fuchsia/yellow No. 2, fuchsia/cupcakes, cinnamon/mechanical, cinnamon/yellow No. 2, cinnamon/cupcakes, tangerine/mechanical, tangerine/yellow No. 2, tangerine/cupcakes, gold/mechanical, gold/yellow No. 2, gold/cupcakes, lime/mechanical, lime/yellow No. 2, lime/cupcakes, forest/mechanical, forest/yellow No. 2, forest/cupcakes, teal/mechanical, teal/yellow No. 2, teal/cupcakes, cobalt/mechanical, cobalt/yellow No. 2, cobalt/cupcakes, plum/mechanical, plum/yellow No. 2, plum/cupcakes}.

(b) The state space has $9 \times 3 = 27$ elements. Of those, two involve a pen whose color begins with "f" and a mechanical pencil. Thus the probability of grabbing a pen whose color begins with "f" and a mechanical pencil is $\frac{2}{27}$.

(c) There are three greenish pens (lime, forest, teal) and two non-mechanical pencils (yellow No. 2, printed with cupcakes) so there are six elements in the state space that comply with the given constraints. Thus the probability of grabbing a pen whose color is greenish and a non-mechanical pencil is $\frac{6}{27} = \frac{2}{9}$.

(d) *Approach the first:* In addition to the six possibilities for (grabbing a pen whose color is greenish and a non-mechanical pencil), there are two possibilities for (grabbing a pen whose color begins with "f" and a non-mechanical pencil) AND NOT (grabbing a pen whose color is greenish and a non-mechanical pencil). Thus the probability is $\frac{8}{27}$.

Approach the second: Let the event of grabbing a pen whose color is greenish and a non-mechanical pencil be G and let the event of grabbing a pen whose color begins with "f" and a non-mechanical pencil be F. Then we seek $P(G \text{ or } F) = P(G) + P(F) - P(G \text{ and } F)$. There are four possibilities for F. There are two possibilities for grabbing a pen whose color (is greenish and begins with "f") and a non-mechanical pencil, so we have $\frac{6}{27} + \frac{4}{27} - \frac{2}{27} = \frac{8}{27}$.

3. A computer lab has 20 computers in it. On any given day, the probability that a given computer is not working is p. How many computers do you expect will be functioning when you enter the lab today? Answer the question for $p = .001, p = .05$.

Approach 1: We will compute the number of computers we expect to be broken and subtract from 20. Let the random variable C count the number of computers that are broken on a given day. We can write $C = \sum_{j=1}^{20} C_j$, where

$$C_j(\text{day } d) = \begin{cases} 1 & \text{computer } C_j \text{ is not working,} \\ 0 & \text{computer } C_j \text{ is working.} \end{cases}$$

$P(C_j = 1) = p$, so $\mathbb{E}[C_j] = 1P(C_j = 1) + 0P(C_j = 0) = p$. Then, by Theorem 14.7.1, $\mathbb{E}[C] = \sum_{i=1}^{20} \mathbb{E}[C_j] = \sum_{k=1}^{20} p = 20p$ broken computers. Thus we expect $20 - 20p$ computers to be functioning on any given day.

Approach 2: We will compute directly the number of computers we expect to be functioning. Let the random variable C count the number of computers that are working on a given day. We can write $C = \sum_{j=1}^{20} C_j$, where

$$C_j(\text{day } d) = \begin{cases} 1 & \text{computer } C_j \text{ is working,} \\ 0 & \text{computer } C_j \text{ is not working.} \end{cases}$$

$P(C_j = 1) = 1 - p$, so $\mathbb{E}[C_j] = 1P(C_j = 1) + 0P(C_j = 0) = 1 - p$. Then, by Theorem 14.7.1, $\mathbb{E}[C] = \sum_{i=1}^{20} \mathbb{E}[C_j] = \sum_{k=1}^{20}(1 - p) = 20 - 20p$ working computers. Thus we expect $20 - 20p$ computers to be functioning on any given day.

When $p = .001$, we expect $20 - 20(.001) \approx 20$ computers to be functioning on any given day.
When $p = .05$, we expect $20 - 20(.05) \approx 19$ computers to be functioning on any given day.

4. Chips of the World come in lots of flavors. In the sale bin are
   2 bags of bacon ranch pita chips,
   1 bag of salt and vinegar potato chips,
   3 bags of hot-sauce cheese corn chips,
   5 bags of crab potato chips, and
   2 bags of peppercorn salsa pita chips.
   If you close your eyes and grab three bags of chips (one at a time, so you know you have three), what is the probability you will get
   . . .

   🦆 . . . all three bags of potato chips?

   🦆 . . . exactly 2 bags of spicy chips?

   🦆 . . . at least one bag of pita chips?

How many bags of corn chips do you expect to find in your three bags?

All three potato: There are a total of 6 bags of potato chips out of 13 bags of chips. So the probability of the first bag grabbed being potato is $\frac{6}{13}$. Twelve bags remain, of which 5 are potato, and then 11 bags remain of which 4 are potato, so the total probability is $\frac{6}{13} \cdot \frac{5}{12} \cdot \frac{4}{11}$.

Exactly two spicy: What counts as spicy chips? Probably hot-sauce cheese corn chips and peppercorn salsa pita chips, for a total of 5 of the 13 bags of chips. There are three ways that we could get exactly two spicy-chip bags—the first, or the second, or the third bag isn't of spicy chips. So we have $\frac{8}{13} \cdot \frac{5}{12} \cdot \frac{4}{11} + \frac{5}{13} \cdot \frac{8}{12} \cdot \frac{4}{11} + \frac{5}{13} \cdot \frac{4}{12} \cdot \frac{8}{11}$. Notice that the set of numerators and denominators is the same in each case, so this is equal to $3 \cdot \frac{8 \cdot 5 \cdot 4}{13 \cdot 12 \cdot 11}$.

At least one pita: To get at least one bag of pita chips, we could add the probability of getting exactly one bag of pita chips (three ways to do that) to the probability of getting exactly two bags of pita chips (three ways to do that) to the probability of getting exactly three bags of pita chips (just one way to do that). Or we could observe that (probability of grabbing at least one bag of pita chips) $= 1 -$ (probability of grabbing no bags of pita chips). It's easy to compute the probability of grabbing no bags of pita chips—there are 3 bags of pita chips out of the 13 bags, so we just get $\frac{10}{13} \cdot \frac{9}{12} \cdot \frac{8}{11}$.

Number of corn: To compute the number of bags of corn chips we expect to find in our three bags, we need to compute expected value, which means we need to define a random variable. Let's let C count the number of bags of corn chips we find. By definition, $\mathbb{E}[C] = 0P(C = 0) + 1P(C = 1) + 2P(C = 2) + 3P(C = 3)$. So, we need to compute these probabilities. There are 3 bags of corn chips among our 13.

$P(C = 0) = \frac{10}{13} \cdot \frac{9}{12} \cdot \frac{8}{11}$. (Okay, that wasn't necessary. Too late.)

$P(C = 1) = 3 \cdot \frac{3 \cdot 10 \cdot 9}{13 \cdot 12 \cdot 11}$.

$P(C = 2) = 3 \cdot \frac{3 \cdot 2 \cdot 10}{13 \cdot 12 \cdot 11}$.

$P(C = 3) = \frac{3}{13} \cdot \frac{2}{12} \cdot \frac{1}{11}$.

Thus $\mathbb{E}[C] = 3 \cdot \frac{3 \cdot 10 \cdot 9}{13 \cdot 12 \cdot 11} + 2 \cdot 3 \cdot \frac{3 \cdot 2 \cdot 10}{13 \cdot 12 \cdot 11} + 3 \cdot \frac{3}{13} \cdot \frac{2}{12} \cdot \frac{1}{11} = \frac{9}{13}$ expected bags of corn chips.

5. Shoes 'R' Us has a lot of different kinds of shoes in their display case, one of each kind they sell. A shoe can be brown, black, silver, or green; it can be a low shoe, a boot, or an athletic shoe; and, it can have laces or be a slip-on.

   (a) How many different kinds of shoes does Shoes 'R' Us have in its display case?

   (b) What is the probability that a Shoes 'R' Us display shoe is brown and slip-on?

   (c) What is the probability that a Shoes 'R' Us display shoe is silver or a boot?

   (d) Given that a Shoes 'R' Us display shoe is silver, what is the probability that it is a boot?

   (e) Given that a Shoes 'R' Us display shoe is a green athletic shoe, what is the probability that it has laces?

   (f) Are the properties *silver* and *boot* independent?

(a) Shoes 'R' Us shows $4 \cdot 3 \cdot 2 = 24$ kinds of shoes in its display case.

(b) Of those, 6 are brown and of the brown shoes, 3 are slip-on. So the probability is $\frac{3}{24} = \frac{1}{8}$.

(c) There are 6 silver shoes in the display, and 8 boots in the display case, and (check it out, we're about to use PIE) two of those are silver boots. Therefore, there are $6 + 8 - 2 = 12$ display shoes that are silver or boots, and the probability is $\frac{12}{24} = \frac{1}{2}$.

(d) There are 6 silver shoes, of which 2 are boots, so the probability of a silver shoe being a boot is $\frac{1}{3}$. We could also use the conditional probability formula and get $\frac{1/12}{1/4} = \frac{1}{3}$.

(e) There are 6 green shoes, of which 2 are also athletic. One of the green athletic shoes has laces, so the probability is $\frac{1}{2}$.

(f) To check for independence of *silver* and *boot*, we check to see whether $P(boot|silver) = P(boot)$. We already know that $P(boot|silver) = \frac{1}{3}$. And $P(boot) = \frac{1}{3}$, so these properties are independent.

6. Consider a deck of cards that is standard, except for having six suits—the two additional suits are stars and squids. (This deck exists: it is the Blue Sea Deck.) Draw a card.

(a) What is the probability that the card is a queen or a squid?

(b) What is the expected value of the number on the card? (Here, Ace $= 1$, King $= 13$.)

(a) In this deck, there are 6 queens. And there are 13 squids. There is a queen-of-squids card, so the total number of queens or squids is (by PIE) $6 + 13 - 1 = 18$. The total number of cards is $6 \cdot 13 = 78$. Therefore, the probability of a card being a queen or a squid is $\frac{18}{78} = \frac{3}{13}$.

(b) We define the random variable X to measure the value of the number on a card. Let's use Lemma 14.3.10 to calculate the expected value. There are 78 cards, and so 78 possible states in the space S. Each of these states has a probability of $\frac{1}{78}$, so

$$\mathbb{E}[X] = \sum_{s \in S} X(s)P(s) = \sum_{s \in S} X(s) \cdot \frac{1}{78} = \frac{1}{78} \sum_{s \in S} X(s). \text{ The values } X(s)$$

can have range from 1 to 13, and there are 6 of each, so we now have $\frac{1}{78} \cdot 6(1 + \cdots + 13) = \frac{1}{78} \cdot 6 \cdot 91 = 7$.

7. The game *Elder Sign* has unusual dice. There are six 6-sided green dice, each of which has three sides showing magnifying glasses, one side with a tentacle, one side with a skull, and one side with a scroll. There is also a 6-sided yellow die with four sides showing magnifying glasses, one side with a skull, and one side with a scroll. Finally, there is a 6-sided red die with three sides showing magnifying glasses, one side with a Wild sign, one side with a skull, and one side with a scroll.

   (a) If you roll the six green dice, what is the expected number of magnifying glasses you'll see?

   (b) If you roll all seven of the dice, what is the probability that you will roll exactly one skull?

   (c) If you roll all seven of the dice, what is the probability that you will roll at least one scroll?

(a) We define the random variable M to be the number of magnifying glasses we see when rolling the six green dice. This computation will be a lot easier if we define M_i to be 1 if we get a magnifying glass on the ith die, and 0 if we don't—we can note that $M = \sum_{i=1}^{6} M_i$ and use Theorem 14.7.1.
$\mathbb{E}[M_i] = 0 + P(M_i = 1) = \frac{3}{6} = \frac{1}{2}$. Therefore,
$\mathbb{E}[M] = \sum_{i=1}^{6} \mathbb{E}[M_i] = 6\mathbb{E}[M_i] = \frac{6}{2} = 3$.

(b) The rolls of the dice are all independent of each other, and each die has exactly one skull, and there are 7 ways to choose the die that shows a skull, so the probability is $7 \cdot \frac{1}{6} \left(\frac{5}{6}\right)^6 \approx .39$.

(c) Observe that (probability of rolling at least one scroll) $= 1 -$ (probability of rolling no scrolls), so we compute $1 - \left(\frac{5}{6}\right)^7 \approx .72$.

8. The game of Qwirkle uses a bag of tiles. Each black tile has a shape on it (circle, diamond, square, crisscross, starburst, clover) that is colored (red, orange, yellow, green, blue, purple). There are three copies of each kind of tile.

   (a) How many tiles are in a Qwirkle bag?

   (b) What is the probability that a tile drawn is red?

   (c) What is the probability that a tile drawn is a sunburst?

   (d) What is the probability that a tile drawn is a red sunburst?

   (e) What is the probability that a tile drawn is red or a sunburst?

   (f) Is red-ness independent of sunburst-ness?

(a) There are $6 \cdot 6 \cdot 3 = 108$ tiles in a Qwirkle bag.

(b) Of the tiles, $\frac{1}{6}$ are red, so the probability of drawing a red tile is $\frac{1}{6}$.

(c) Of the tiles, $\frac{1}{6}$ are sunbursts, so the probability of drawing a sunburst tile is $\frac{1}{6}$.

(d) Of the red tiles, $\frac{1}{6}$ are sunbursts, so the probability of drawing a red sunburst tile is $\frac{1}{36}$.

(e) There are 18 red tiles and 18 sunburst tiles, and 3 red sunburst tiles, so (PIE!) there are $18 + 18 - 3 = 33$ tiles that are red or sunburst. Thus the probability of drawing a red or sunburst tile is $\frac{33}{108} = \frac{11}{36}$.

(f) To check independence, we'll note that the conditional probability of red-given-sunburst is equal to that of sunburst-given-red (both are $\frac{1/36}{1/6} = \frac{1}{6}$) and equal to that of sunburst and of red (each $\frac{1}{6}$), so these events are independent.

9. Another Qwirkle qwestion: Pull two tiles from the bag.

   (a) What is the probability that both are blue?

(b) What is the probability that the *second* tile is blue?

(c) What is the probability that at least one tile is blue?

(a) We draw tiles one at a time; the first has a probability of $\frac{36}{108}$ of being blue and the second of $\frac{35}{107}$ (because only 107 tiles are left when we draw the second tile). So the probability of drawing two blue tiles is $\frac{36}{108} \cdot \frac{35}{107}$.

(b) If we draw two tiles and the second is blue, we either had first-tile-blue and second-tile-blue (probability $\frac{36}{108} \cdot \frac{35}{107}$), or first-tile-not-blue and second-tile-blue (probability $\frac{72}{108} \cdot \frac{36}{107}$). These events are exclusive, so we add their probabilities to obtain $\frac{36}{108} \cdot \frac{35}{107} + \frac{72}{108} \cdot \frac{36}{107}$.

(c) We could compute the probability of getting only the first tile blue and add that to the probability of getting only the second tile blue and add that to the probability of getting both tiles blue. Or we could compute $1 - $ (probability of getting no blue tiles). We already know the probability of getting only the second tile blue ($\frac{72}{108} \cdot \frac{36}{107}$) and the probability of getting both tiles blue ($\frac{36}{108} \cdot \frac{35}{107}$), so we might as well just add in the probability of getting only the first tile blue ($\frac{36}{108} \cdot \frac{72}{107}$) for a total of $2 \cdot \frac{72 \cdot 36}{108 \cdot 107} + \frac{36}{108} \cdot \frac{35}{107} \approx .56$.

10. What's the expected number of fixed points (items that do not move) in a permutation of n items?

We will let F count the number of fixed points in a permutation. For each of the n items, we define a random variable F_i that evaluates to 1 if the item is fixed and 0 if the item moves. There are n possibilities for where the ith item goes in a permutation, so the probability of an item staying in the same place is $\frac{1}{n}$. Thus, $\mathbb{E}[F_i] = 1 P(F_i = 1) + 0 P(F_i = 0) = \frac{1}{n}$. Then, by Theorem 14.7.1, $\mathbb{E}[F] = \sum_{i=1}^{n} \mathbb{E}[F_i] = \sum_{k=1}^{n} \frac{1}{n} = n\frac{1}{n} = 1$ fixed point.

Hey, that seems an awful lot like the example at the end of Section 14.7 in DMwD ... hmmm ...

Chapter
15 🦆🦆🦆🦆🦆🦆🦆🦆🦆🦆🦆🦆🦆🦆🦆

Fun with Cardinality

We define the cardinality of infinite sets and learn how to compare the sizes of various infinite sets.

15.1 Chapter 15 Definitions and Notation

15.1.1 Notation

$|A|$: The cardinality, or number of elements in, a set A.

$\aleph_0$: The cardinality of $\mathbb{N}$.

$\aleph_1$: The smallest number that is bigger than $\aleph_0$.

$\bigcup_{k \in \mathbb{N}} A_k$: The union of countably many sets.

$\bigcup_{\alpha \in I} \Lambda_\alpha$: The union of uncountably many sets (here I is an uncountable index set such as $[0, 1]$).

ZFC: The axiom set we usually use for mathematics, namely the Zermelo-Fraenkel axioms plus the axiom of choice.

15.1.2 Definitions

cardinality: The number of elements in a set.

same size: We consider two sets to be the same size if we can put them in one-to-one correspondence.

aleph-naught: $\aleph_0$, also pronounced *aleph-null*, the cardinality of the natural numbers.

countable: Any set with size $\aleph_0$ (because we can count the natural numbers).

uncountable: Any set with size larger than $\aleph_0$.

Zermelo-Fraenkel axioms: The set of axioms underlying ordinary mathematics. Generally used with the additional axiom of choice.

continuum hypothesis: The statement that $2^{\aleph_0} = \aleph_1$ is known as the continuum hypothesis because it says that the cardinality of the real numbers, or continuum, is $\aleph_1$.

15.2   Chapter 15 Facts and Theorems

Central fact.   If there is a bijective map from A to B, then $|A| = |B|$. Therefore, we can use bijections to determine cardinality for infinite sets. However, to prove that $|A| = |B|$ we do not necessarily need to give a single function that we can prove is a bijection. We can instead show that a set of functions together provide an injection and a surjection.

Ambiguous language.   We have to be very precise when speaking about infinity. For example, we could colloquially say that $\mathbb{N}$ and $2\mathbb{N}$ are the same size (because they have the same cardinality) and we could also say that $\mathbb{N}$ is larger than $2\mathbb{N}$ (because $\mathbb{N} \supset 2\mathbb{N}$ and $\mathbb{N}$ has *way* more elements). But it doesn't make sense for one set to be larger than another set that is the same size!

The continuum hypothesis situation.   The continuum hypothesis states that $2^{\aleph_0} = \aleph_1$. It cannot be shown that this statement is either true or false: It has been proven that under ZFC, if you assume that $2^{\aleph_0} = \aleph_1$, no contradiction arises. It has *also* been proven that if you assume that $2^{\aleph_0} \neq \aleph_1$, no contradiction arises!

On the other hand, Hugh Woodin proposed (in 2001) adding an axiom to ZFC that would make the continuum hypothesis false. In other words, with his axiom, assuming that $2^{\aleph_0} = \aleph_1$ would lead to a contradiction.

15.3   Some Straightforward Examples of Chapter 15 Ideas

An example of countable vs. uncountable.   Our three favorite sets $\mathbb{N}, \mathbb{Z}$, and $\mathbb{Q}$ are countable. We show a set is countable by exhibiting a bijection with $\mathbb{N}$.

On the other hand, $\mathbb{R}$, $[0,1]$ and $\mathscr{P}(\mathbb{N})$ are uncountable. To show a set is uncountable, we show by contradiction that there cannot be a one-to-one correspondence with $\mathbb{N}$.

An example of bijections using multiple maps.   Let us show that $|\mathbb{N} \cup \mathbb{N} \cup \mathbb{N} \cup \mathbb{N} \cup \mathbb{N}| = |\mathbb{N}|$.

First, some setup. We will denote an element $t \in \mathbb{N} \cup \mathbb{N} \cup \mathbb{N} \cup \mathbb{N} \cup \mathbb{N}$ by $t = (n, p)$ where $n \in \mathbb{N}$ and $p \in \{1,2,3,4,5\}$ to indicate which copy of $\mathbb{N}$ we're looking at. Next, we partition $\mathbb{N}$ into five subsets,
$\{n \mid n \equiv 0 \pmod 5\}$, $\{n \mid n \equiv 1 \pmod 5\}$, $\{n \mid n \equiv 2 \pmod 5\}$,
$\{n \mid n \equiv 3 \pmod 5\}$, $\{n \mid n \equiv 4 \pmod 5\}$.

Now we will define our bijection. $f(t) = f((n,p)) = 5(n-1) + p$.
It is an injection: Suppose
$f(t_1) = f(t_2)$. Then we have
$5(n_1 - 1) + p_1 = 5(n_2 - 1) + p_2$.
Examining this equation $\pmod 5$, we see that $p_1 \equiv p_2 \pmod 5$.
Because $p_1, p_2 \in \{1,2,3,4,5\}$, we know $p_1 = p_2$.
Then $5(n_1 - 1 = 5(n_2 - 1)$ implies $n_1 = n_2$.
Therefore, $(n_1, p_1) = (n_2, p_2)$.
It is a surjection: For any $m \in \mathbb{N}$, we note that $m = 5k + r$ for some $k \in \mathbb{W}$ and $r \in \{0,1,2,3,4\}$, so $f((k+1,r)) = 5(k+1-1) + r = m$.

15.4   More Problems for Chapter 15

Those solutions that model a formal write-up (such as one might hand in for homework) are to Problems 4, 5, and 8.

1. Show that $|\mathbb{Z}| = |\mathbb{Z}| + 72$.

2. Show that $\mathbb{Z}$ has the same cardinality as $4\mathbb{N}$.

3. Show that $\mathbb{Z}$ has the same cardinality as $N \times N$.

4. Prove that $|\mathscr{P}(\mathbb{Q})| > |\mathbb{Q}|$.

5. Show that $\mathbb{W}$ has the same cardinality as $\mathbb{Z}$.

6. What is the cardinality of the set $\{\frac{p}{q} \mid p \in \mathbb{W}, q \in \mathbb{Z}\}$?

7. What is $(\aleph_0)^3$? How about $(\aleph_0)^8$? Or $(\aleph_0)^{\aleph_0}$? Explain.

8. Consider the set $\mathscr{F}$ of all functions from $\mathbb{N}$ to $\mathbb{N}$. Is $\mathscr{F}$ countable or uncountable?

9. Is the total number of steps in an algorithm that does not terminate countable or uncountable?

10. Consider the set H of length-$\frac{1}{2}$ intervals that are contained in the interval $[0, 1]$. What is $|H|$?

15.5 More Solutions for Chapter 15

1. Show that $|\mathbb{Z}| = |\mathbb{Z}| + 72$.

   We will exhibit a bijection between $\mathbb{Z}$ and $\mathbb{Z} \cup \mathbb{Z}_{72}$. For $k \leq 0$, let
   $f(k) = k \in \mathbb{Z}$, for $1 \leq k \leq 72$ let $f(k) = k - 1 \in \mathbb{Z}_{72}$, and for $k > 72$,
   let $f(k) = k - 72 \in \mathbb{Z}$.
   This is injective: Note that if $f(k) = f(r)$, then $f(k), f(r)$ must be
   both in $\mathbb{Z}$ or both in $\mathbb{Z}_{72}$. If $f(k) = f(r)$, then both are positive or
   both are nonpositive. If both are nonpositive, then $k = r$. If both are
   positive, then either $k - 72 = r - 72$, so $k = r$, or $k - 1 = r - 1$, so
   $k = r$.
   It is also surjective: for $z \in \mathbb{Z}$, if $z \leq 0, f(z) = z$, and if $z > 0$,
   $f(z + 72) = n$ and for $z \in \mathbb{Z}_{72}, f(z + 1) = z$.

2. Show that $\mathbb{Z}$ has the same cardinality as $4\mathbb{N}$.

   We will exhibit a bijection between $\mathbb{Z}$ and $4\mathbb{N}$. For $k < 0$, let $f(k) =$
   $-8k$. For $k \geq 0$, let $f(k) = 8k + 4$.
   Injective: Suppose $f(k) = f(r)$. Both must be divisible by 8 or
   not divisible by 8. If $f(k) = f(r)$ is divisible by 8, then we have
   $-8k = -8r$ so $k = r$. If $f(k) = f(r)$ is not divisible by 8, then we
   have $8k + 4 = 8r + 4$ so $k = r$.
   Surjective: Every element n of $4\mathbb{N}$ is divisible by 4; and, either divisible by 8, or $\equiv 4 \pmod{8}$. If n is divisible by 8, then $f(-\frac{n}{8}) = n$
   and if n is not divisible by 8, then $f(\frac{n-4}{8}) = n$.

3. Show that $\mathbb{Z}$ has the same cardinality as $N \times N$.

   We will exhibit a bijection between $\mathbb{Z}$ and $N \times N$. We associate
   $0 \leftrightarrow (1,1), 1 \leftrightarrow (1,2), -1 \leftrightarrow (2,1)$, and continue in the pattern
   indicated by Figure 15.1.

4. Prove that $|\mathscr{P}(\mathbb{Q})| > |\mathbb{Q}|$.

   Proof by contradiction: Suppose that there is a bijection f between
   $\mathscr{P}(\mathbb{Q})$ and $\mathbb{Q}$. Consider the set $Q \in \mathscr{P}(\mathbb{Q})$ defined as $Q = \{q \mid q \notin$
   Q', where Q' is such that $f(Q') = q'\}$. Consider $f(Q)$. Is $q \in f(Q)$?
   Suppose $q \in f(Q)$. By definition, $q \notin f(Q)$, which is a contradiction.

5. Show that $\mathbb{W}$ has the same cardinality as $\mathbb{Z}$.

   We will exhibit a bijection between $\mathbb{W}$ and $\mathbb{Z}$. When $w \in W$ is odd,
   let $f(w) = \frac{w-1}{2} + 1$ and when $w \in W$ is even, let $f(w) = \frac{-w}{2}$.

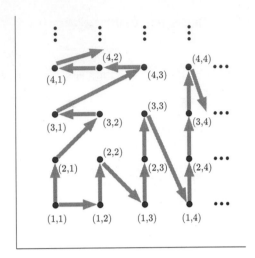

Figure 15.1. A bijection between $\mathbb{Z}$ and $N \times N$.

This is injective: If $f(k) = f(r)$, then either both values are positive, so that $\frac{k-1}{2} + 1 = \frac{r-1}{2} + 1$, so $k = r$, or both values are non-positive, so $\frac{-k}{2} = \frac{-k}{2}$ so $k = r$.

It is also surjective: for $n \in \mathbb{Z}$, if $n > 0$, then $f(2(n-1)+1) = \frac{2(n-1)+1-1}{2} + 1 = n$ and if $n \leq 0$, then $f(-2n) = \frac{-(-2n)}{2} = n$.

6. What is the cardinality of the set $\{\frac{p}{q} \mid p \in \mathbb{W}, q \in \mathbb{Z}\}$?

    There is a bijection between this set and $\mathbb{W} \times \mathbb{Z}$, defined by $f(\frac{p}{q}) = (p, q)$. We know that $|\mathbb{W}| = |\mathbb{Z}| = |\mathbb{N}| = |\mathbb{N} \times \mathbb{N}|$. Therefore, the cardinality is $\aleph_0$.

7. What is $(\aleph_0)^3$? How about $(\aleph_0)^8$? Or $(\aleph_0)^{\aleph_0}$? Explain.

    All three of these are equal to $\aleph_0$. For example, $(\aleph_0)^3 = \aleph_0 \cdot \aleph_0 \cdot \aleph_0 = |\mathbb{N} \times \mathbb{N} \times \mathbb{N}|$, and we know from Section 15.5 of the text that there is a bijection between $\mathbb{N} \times \mathbb{N} \times \mathbb{N}$ and $\mathbb{N}$. The same holds for $(\aleph_0)^8$, and even $(\aleph_0)^{\aleph_0}$ because there is a bijection between $\bigcup_{k \in \mathbb{N}} \mathbb{N}_k$ and $\mathbb{N}$.

8. Consider the set $\mathscr{F}$ of all functions from $\mathbb{N}$ to $\mathbb{N}$. Is $\mathscr{F}$ countable or uncountable?

    We will show that $\mathscr{F}$ is uncountable by showing that a proper subset of $\mathscr{F}$ is uncountable. Consider a subset $A \subset \mathbb{N}$. There exists a function $f_A : \mathbb{N} \to \mathbb{N}$ defined by

$$f_A(n) = \left\{ \begin{array}{ll} n & n \in A, \\ 1 & n \notin A. \end{array} \right.$$

In other words, f_A sends elements of A to themselves and sends everything else to 1. (This is neither one-to-one nor onto unless $A = \mathbb{N}$.) And, the functions f_A are in one-to-one correspondence with elements $A \in \mathscr{P}(A)$, so $|\mathscr{F}| \geq |\mathscr{P}(A)|$, and we know $\mathscr{P}(A)$ is uncountable because $|\mathscr{P}(A)| = 2^{\aleph_0} > \aleph_0$.

9. Is the total number of steps in an algorithm that does not terminate countable or uncountable?

   The steps in an algorithm are numbered, so we can count them as the algorithm proceeds and therefore the total number is countable.

10. Consider the set H of length-$\frac{1}{2}$ intervals that are contained in the interval $[0,1]$. What is $|H|$?

    The intervals of length $\frac{1}{2}$ are in one-to-one correspondence with their left endpoints. (If you don't believe this, consider that given a left endpoint we can find the right endpoint by adding $\frac{1}{2}$.) An interval of length $\frac{1}{2}$ inside the unit interval can have a left endpoint as small as 0 and as large as $\frac{1}{2}$. So $|H| = |[0,\frac{1}{2}]|$. The elements of $[0,\frac{1}{2}]$ are in one-to-one correspondence with the elements of $[0,1]$, so $|H| = \aleph_1$.

Printed and bound by CPI Group (UK) Ltd, Croydon, CR0 4YY

22/10/2024

01777624-0006